Handbook of Submarine Operations

P.R. Franklin

ISBN 13: 978-93-856990-1-6 ISBN 10: 93-856990-1-6

Published by

Frontier India Technology
No 22, 4th Floor, MK Joshi Building, Devi Chowk, Shastri Nagar,
Dombivli West, Maharashtra, India. 421202
http://frontierindia.org
https://www.facebook.com/frontierindiapublishing

DEDICATION

This book is dedicated to all submariners serving in the Indian Navy. May all their dives be followed by surfacing's, and may they always have twenty feet beneath their keel when dived.

CONTENTS

ACKNOWLEDGMENTS

I am indebted to all my Instructors in the erstwhile Soviet Union who ingrained the art of operating submarines in me; to all my fellow submariners, both senior and junior, who enriched my professional knowledge, and with whom I had the pleasure to serve all through my service in the Submarine Arm of the Indian Navy. National Maritime Foundation, New Delhi, had published the first edition of this book.

PREFACE

"The Submarine has created its own type of officer and man with language and traditions apart from the rest of the service, and yet at heart unchangingly of the service."

Rudyard Kipling
"The Fringes of the Fleet, 1915".

They first fought on land. With the discovery that they could float on water, borne by catamarans, rafts, crafts and the likes, they next fought on water. The Wright brothers ushered in the era of flight and soon they fought in the air also. Fighting underwater was more complicated. However by 1864, with the introduction of the submarine as a successful fighting platform (during the war of the States, the spar-torpedo-armed submarine "H.L.Hunley" sank the Federal navy's sloop, the "Housatonic" on the night of 17 Feb.1864), they began to wage war from underwater too. Man is now itching to carry war into space, and the day is not far when that too becomes a reality. Wars were, are, and will continue to be, a necessary dark side of humankind.

Undersea warfare or 'Inner Space' warfare as some like to call it, developed and progressed through the two Great Wars and the Cold War that followed. From operating on the surface and diving only to attack, submarines have developed into platforms that can now stay 'indefinitely' under water and hunt not only ships and other submarines, but land targets too. Only human endurance limits the underwater sojourn. Living and working in a steel capsule underwater does take its toll on man.

Navies have stringent specifications for operating submarines and those who man them. The psychological demands made of individuals in submarine service dictate that they be volunteers. Submarines all over the world have been, and mostly still are, run by volunteers from the general cadre of their respective navies. Only when navies find it difficult to get sufficient volunteers do they resort to other means of inducting personnel into submarines. Candidates have to undergo a stiff medical examination, and a series of psychological and other tests, the successful completion of which enable them to commence submarine training. The criteria for selection concentrates on their fitness to work in confined spaces in close contact with others for long durations. To successfully work in confined spaces, teamwork, a calm disposition, pleasing social behavior, a pleasant nature, good habits, emotional stability, and an absence of psychopathic behavior is of prime importance. Individuals with suicidal tendencies and those with personality disorders are rejected. Other attributes not desirable in individuals are sub normal intelligence, overt and covert anxiety disorders (claustrophobia, social anxiety, and obsessive compulsive disorders), lack of motivation, history of personal ineffectiveness, bad inter-personal relationships, and a lack of adaptability.

To live and work efficiently for days on end, in claustrophobic surroundings inside a steel capsule crammed with men, nuclear reactors, nuclear and conventional ordnance, other machinery and equipment; in an unnatural and polluted limited air environment resulting from the presence of humans, batteries, other chemicals emitting gases, cooking, machinery operating, and the all-pervasive smells of oils and greases; in artificial lighting day in and day out and night in and night out; and in a world with upsetting and unnatural body cycles and biorhythms, requires a special breed of men. Yet, these are not superhuman men. They look, behave, walk, and talk like any of the men in the rest of the navy. What then propels them to volunteer for this demanding Arm, Service, or Cadre?

Is it the extra remunerations that are normally offered for such service? Some navies do and others don't. Admittedly, some volunteer for this very reason, and accept the consequential hardships. Some: not all. Is it the glamour attached to this branch of Service? Pilots in the aviation cadre, marine commandos, and divers also have this certain aura about their profession, and submariners are not far behind. Badges worn on uniforms by pilots, SEALS, clearance divers, commandos and submariners all catch the eye and respect of those without them. Again, this may be a reason why some volunteer. Is it the challenge of the risk factor that draws these men to volunteer? Humans love challenge and are constantly stretching and testing themselves in various fields, beyond their individual abilities and talents: perhaps. Is it the spirit of adventure that this career offers? Ask a cross section of submariners and one will get different answers. However, all of them will have more than one reason, and partially agree to the reasons just enumerated. Whatever the reasons, service in submarines draws the best out of men who sail out for difficult missions, performing the most arduous of duties demanding the highest standards of discipline and drill, and develops a bond of camaraderie and esprit-de-corps that is hard to find elsewhere. Give them a choice to leave this cadre and you will find reluctance with a capital 'R'. Most of those who have served, and are now too old to serve, will give any number of years to be young again and serve once more in submarines. This is inexplicably true.

What sets these men apart from the rest of the navy? For one, most of these platforms operate as single units out at sea for most of the time. Onboard, the men have opted for a life of living at close quarters demanding rigorous discipline and isolation from the outside world. Each man has specific tasks and responsibilities that he must carry out. At a very early age he learns to do things correctly and work without supervision. The safety and performance of the boat depends on the actions of every man jack onboard. At a very young age, he has to take decisions – and very quick ones at that – all by himself. A breakdown of machinery or systems in his vicinity demands instant corrective action. The decision to opt for one of multiple options has to be the correct and best one. One wrong move or decision can cost heavily. That is a very great responsibility thrust on very young shoulders.

The Captain is also invariably very young. On him rests a huge responsibility at an early age. For him, handling the submarine is predominantly a mental activity – a difficult task of continuously discerning the undersea surroundings. Onboard equipment helps him. He is, however, alone in 'decision- making'. For most of the time, there are no Fleet Commanders or 'higher ups' to consult every now and then. That is the privilege of the surface navy. Here, underwater, he must make his own assessments and take decisions. Wrong moves in peacetime can create international disquiet. When operating singly in a hostile environment, the decision to attack or evade is again a choice that has to be made by him, and him alone. The sequence and selection of targets in a multi-ship environment is again a lone decision. Fleet ships share targets after much consultation between themselves. The submarine Captain does it alone. Should he attack a screen ship first? Or, go for the 'big one' straightaway? Which screen ship should he go for? From which direction should he go for the big one? Despite all possible homework done ashore before setting out, a change in scenario or circumstances demands new appreciations, assessments, and decisions to be taken inside the steel hull, underwater.

Yes, at a very young age a high quality of decision-making is demanded of the submariner, and he trains to make the correct ones. For another, he trains to assess and understand dangerous situations quickly and exercise that certain boldness in 'follow up' actions. It is this boldness that encourages and ensures an offensive spirit, without compromise of safety and lives of the men in the hull. These qualities are not demanded of personnel of similar age on surface ships. That is what sets the submariner apart from the rest of his counterparts in the navy.

Their work and how they should operate at sea is what this book is all about. For the uninitiated, the simplicity of the language, and the explanations that follow, will make it interesting reading. Technical aspects and submarine jargon are predominant by their absence as different navies adopt different terminologies for the same thing. For those who have just joined this cadre, there is enough to understand the overall scenario under which submarines deploy, operate, and perform. 'Bloggers' will also find something in this book to ruminate on.

Finally, there have been many books written on submarine warfare in many parts of the world and at various stages in history. To err on the side of repetition is an easy route to follow. Particular care has been taken to avoid such an option and to make this book different.

THE UNDERSEA ENVIRONMENT

"Seawater Chemistry, Acoustics, Physics, Geography, Topography, and Oceanography are all essential for submariners to master, in order to conduct their many under water missions safely and successfully."

Anon

Calm and serene, sometimes ominous and threatening, the seemingly endless expanses of waters that form the oceans and seas of this planet never cease to thrill mariners. All those attracted to the sea look at nature's awesome creation that covers seventy percent of the globe, with wonderment and mixed feelings. Everyone is aware of the many globe-trotting voyages of discovery in the days of sail; of the navies that ventured beyond known frontiers and fought great battles in pursuit of annexing and colonizing territories; of the large exchange of culture and trade that plied between widely spaced shores; of the two Great Wars that were fought, sending down millions of tonnage of shipping down to Davy Jones' Locker; of the many ships that have been traversing the waters carrying cargo and passengers from country to country for centuries, and still do. The surface of the sea has been both a friend and an adversary to mankind. Very few are aware of what lies under this watery surface, and there are many cogent reasons for this. Mankind has not been able to move about as freely underwater as in the air, in the ionosphere, and in space. What lurks and exists beneath the waves largely remained a mystery that is only slowly and increasingly being unraveled now, in the present era.

Recapitulate for a moment how different landmasses appear in different parts of the world. Think of the high mountain ranges in each of the continents, the rift valleys formed by subsidence and the erosive effects of fast moving rivulets of melting snow that seek each other to conjoin and form rivers that slow down as they reach the plains, drop their sediments, and wind their way into the sea. Think of the dense foliage in other parts, with fauna and flora aplenty, some negotiable and others absolutely impregnable. Think of the vast expanse of flat lands that cover some terrains. Think of active landmasses in the form of hot springs and volcanoes. Think of the myriads of types of living creatures that exist and make themselves known by their visual presence as much as by their noises. The undersea is no different. All these and more exist under the surface of the sea. It is as complex as the landmass that forms only thirty percent of the globe's surface. The same landmass, if immersed in water, will give one some idea of what lurks beneath the waves, with marine life in equal if not more numbers and variety replacing the fauna on dry land.

The sea is noisy, and we are not talking only about the noise of waves lapping the shoreline. The sea is a fascinating subject for acoustic study, the finished product of which will run into some many volumes. Undersea, there are noises covering a whole range of frequencies, far beyond what humans can hear. At one end of the spectrum – the low frequency end – there are whales bellowing across the oceans for hundreds of miles, calling out to others of their community. They can communicate with their kin from the South Atlantic to the North Atlantic Ocean. At the other end of the spectrum are the myriads of tiny little creatures that communicate with each other in very high frequencies across very short ranges. The noisy scenes in shallow waters and continental shelves are different to those in deep waters. Where mighty rivers flow into the seas, the noises are different to that in surrounding areas. Where subterranean volcanic activities and warm water springs abound, a different lot of inhabitants of the region produce their own peculiar noises. There are creatures that live and survive only at great depths just as there are others who can survive only in shallow areas. The undersea is well populated and far from quiet.

Warships, anti-submarine aircraft, anti-submarine helicopters, and hunter killer attack submarines primarily exploit the science of 'Acoustics' to locate submarines at sea. It is the same science that is used by submarines themselves, using the same acoustic spectrum to their advantage to track unsuspecting adversaries, and to feel their way around underwater like a visually challenged person feels his way about in busy streets and confined areas. It is used just as bats use acoustics in the air in lieu of eyes, or like electronic waves used for the same purpose in the atmosphere. Sound is the preferred option because it travels, and can be heard at much greater distances underwater than radio waves. It is not the intention here, to go into any depth on the subject of underwater acoustics. For those interested in the subject, Robert. J. Urick's book titled "Principles of Underwater Sound" may form the ideal launch platform. It is a very vast subject and there is enough literature by other authors also available on the topic. Information can also be gleaned through the internet. However, a brief mention is being made here to draw the attention to the complexities presented by the medium in which undersea warfare is carried out.

Passive acoustic sensors – those that only listen and do not emit any sound – are used to listen for the typical noises of propulsion and machinery emitted by submarines and other craft over the self-noise of the reverberating sea. All anti-submarine forces have passive sensors and use them when conditions for their use are favorable. Submarines use passive sensors underwater all the time. The alternate means of detection are *Active acoustic sensors* – sensors that transmit sound and look for 'echoes' bounced off the hull of a submarine, bearing in mind that they bounce off large fish, underwater pinnacles, subterranean rocks, and the likes, too. Active sensors broadcast their own presence whereas passive ones do not. While submarines use passive sensors to locate their adversaries, they have active sensors, but use them for specific and limited purposes, and when safe or ordered to do so.

In the complex environment that the sea offers, sound does not travel in straight lines or along the intended path, and herein lies much of the problems in locating and tracking submarines, whether by passive or active means. Seawater is not a homogenous medium. The salinity and temperature vary with time, season, and place. The shallower part of the sea that is just below the surface is subjected to diurnal variations. This layer is affected by day-night variations in temperature, and meteorological variations during different times of the year. These variations affect the ray path of a sound source traveling from and to a shallow sensor. What is known as the afternoon effect reduces echo ranges on calm and sunny days around mid-day. The warm water near the surface prematurely bends sound rays downwards, and submarines lurking at shallow depths are not likely to be detected at normal ranges by sensors fitted on the hulls of ships during this part of the day. Detection may occur only when the range is already too near for comfort. At night this phenomenon disappears. A Surface Duct occurs just below the sea surface whenever the processes of stirring by wind and convection, caused by cooling and evaporation of water, takes place. Here sound is trapped near the surface giving extended acoustic ranges at shallow depths, with occasional losses at the fringes. This mixed layer, or surface duct, is predominant in windy 'Temperate Regions' of the world's oceans. The thickness varies with the seasons. For example, in the temperate regions of the North Atlantic Ocean (between Lat 40° – 50°N) the median of the surface duct is at about 200ft depth in winter (Jan-Mar), 90ft depth in spring (Apr-Jun), 70ft depth in summer (Jul-Sep), and 150ft depth in autumn (Oct-Dec). Sometimes, the presence of a slight negative gradient in this mixed layer prevents a surface duct from forming altogether!

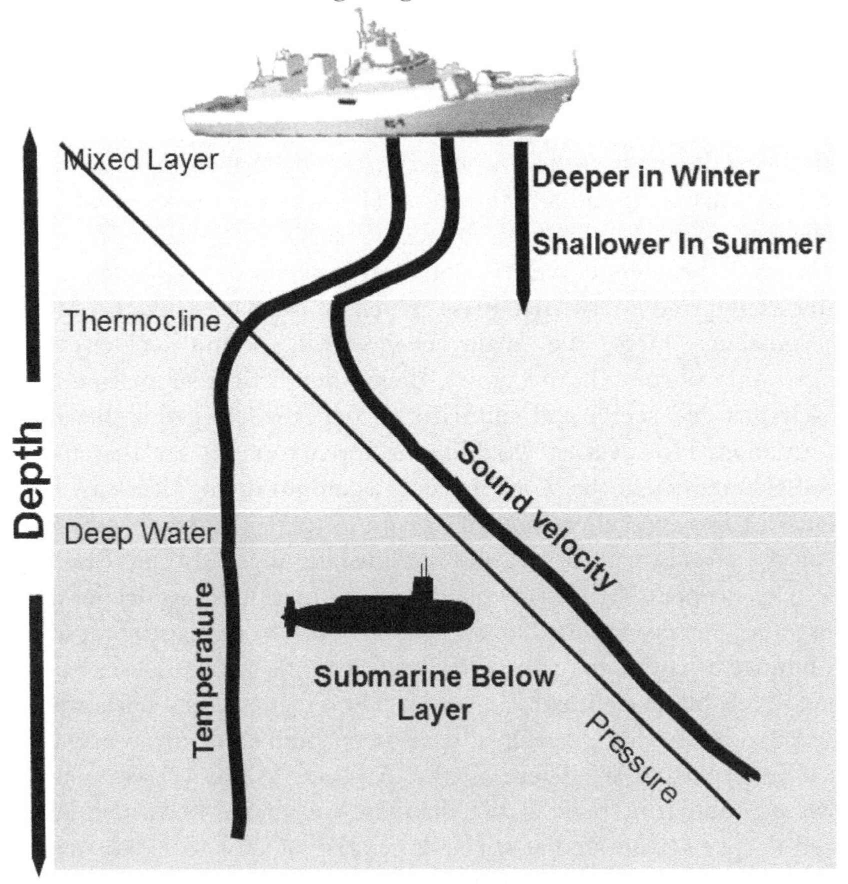

The sea surface affects shallow water acoustics through a reflection loss of sound incident upon it, produces back-scattering which results in sea surface reverberations; produces forward scattering resulting in reduced ranges of detection; casts a shadow in a negative gradient which can extend right up to the surface in which a lurking submarine may not be detected; creates a layer of bubbles of air just below the surface when the sea is rough which results in the sea surface being acoustically hidden; creates ambient noises of a disturbing nature in frequencies between 0.5 to 50 kilohertz, and complicates any attempts to predict the sound ray path with some accuracy in a selected region. The picture gets more complex when moving away from deep oceans to shallow waters. In shallow waters, like for example above continental shelves or near the coast, transmission losses near the sea surface tend to be higher in summers than in winters, and reverberation effects are high.

The *velocity of sound* in water is also dependent on depth or pressure (and pressure is a function of density and temperature). Seasonal variations of heating and cooling penetrate a little deeper and affects the cross section of water below the shallow diurnal layer, down to a depth of about 200 meters. This water mass lies in what is referred to as the *Seasonal Thermocline* bracket.

From 210 meters down to about 1000 meters or so, the mass of water is referred to as that mass which is bracketed by the Main Thermocline. Here the main changes in sound velocity or temperature occur. The maximum diving depth of a submarine is a closely guarded secret and submariners never disclose this piece of information. However, it wouldn't be too wrong to say that most naval submarines in the world have a maximum diving depth of 500 meters or less, and therefore do not go deeper. The water pressure column increases with depth, and the crushing force also increases as one goes deeper. The cost of building a submarine to go deeper and resist the increasing crushing force also becomes exorbitant and prohibitive. Most of the anti-submarine weapons are also designed to operate reliably in depths of less than 500 meters. A few underwater research vessels that go deeper have been built over the years, but have remained 'few'. There is the Russian 'Alpha' class nuclear powered submarine built with a titanium hull that is known to have dived deeper and moved at speeds in excess of 40 knots. Not many were built, as they proved to be too expensive to build. However, ships and submarines do use the deeper seawater medium for their '*sonars*' to locate their adversaries through bouncing sound waves off the bottom of the sea using *bottom bounce* techniques, and also for long range communications through *deep sound channels* (some leakages occur if frequencies less than 20 hertz are used). Deep Sound Channels are caused when the deep sea is warm on top and cold below. A reversal of the velocity gradient occurs, and this happens at a depth of around 1200 meters and results in a reduction in sound velocity, which has an effect on long range sound propagation (Deep Sound Channel). During World War II, a SOFAR (sound fixing and ranging) net was established in the Pacific Ocean by the Americans, which for many years after the war remained the best non-radio channel for long distance communications (satellite communications and internet apart). Beyond this depth the temperature of the water is about 4° C (isothermal), the density is highest, and the sound velocity increases with depth.

The classifications just described are very general and not applicable across the board. For example, as we move from higher latitudes towards the Poles where low temperatures are predominant for the major part of the year, the *Main and Seasonal Thermocline* become less prominent and the zone of reversal of sound velocity, or the Sound Channel, comes shallower till it virtually lies just under the ice in the Arctic. In the Arctic, because of the absence of the main thermocline (present in 'Temperate' latitudes) due to lack of solar heating, acoustic rays present a positive gradient up to shallow depths in summers and go all the way up to the surface ice in winters. In summers, in open waters, acoustic rays often form a slight negative gradient down to 100 feet or more. However, salinity gradients occur where the ice is melting or where rivers are flowing, and this presents a positive gradient, which might neutralize the negative gradients, if existing. The Arctic therefore has some of the characteristics of both the mixed layer channels and the Deep Sound Channels. In the Mediterranean Sea a shallower sound channel exists in the summer months at a depth of around 100 meters with an axial temperature of around 13° Centigrade.

Apart from variations in temperature, pressure, and salinity affecting the velocity of sound in seawater, when energy is radiated by a sound source, there is a loss of energy experienced through spreading and attenuation. There is no regular way of describing this phenomenon because of differences in different oceans. For example, the Pacific Ocean is more acidic than the Atlantic Ocean and therefore attenuation in the Pacific has been observed to be around half that in the Atlantic, for a given frequency.

One may well ask why all this is necessary to know in the conduct of submarine operations and undersea warfare? To locate the exact position of a moving submarine underwater, an *accurate bearing, accurate range*, and, if possible, *accurate depth*, of that platform is required by the locator. Periodic readings need to be recorded to work out its Course and Speed in order to track it or predict its future position as required. Similarly, for a submarine to locate the exact position of a ship, accuracy of bearings is equally vital. Velocity being a function of distance and time, accurate sound velocities is required whenever a *travel time* in the sea has to be converted into *distance*. Because of the effects of the sea on sound rays, the curving pattern of sound rays in a given area of interest is also required, and predictions are called for, for a better understanding of how they will behave. This then gives the *Sound Ray Predictions* (SRP) and the *Expected Sonar Range* (ESR) of a given sensor in a given area during a particular time-period. *Fire Control Systems* onboard submarines require this information where, apart from the target's Course and Speed, an accurate target Range is needed from sonar data, to launch a weapon and expect a high probability of hit or kill. Submarines have a distinct advantage over ships and anti-submarine aircraft in getting up-to-date predictions of behavior of sound in a given region, as they are able to maneuver in the vertical plane, physically move through the vertical zones of interest, and get their SRP and ESR at any given moment quite accurately. Ships and aircraft have to lower sensors vertically down into the water to the required depth to get this information, for which they have to stop, hover (helicopters), or drop sensors in the water (aircraft). These cannot be continuous efforts and therefore their predictions are not likely to be as up to date as they would like them to be. Advantage – submarine!

Navigation underwater presents its own problems to submarines. A submarine has no visual aids to assist it in navigating below Periscope Depth, underwater. There are no lighthouses, no visual aids, no stars or celestial objects, no satellites, that are routinely used by ships for 'position fixing' to navigate on the surface of the sea. Yet, with underwater pinnacles and mountains, currents and upheavals, outflows of rivers and man-made objects like off shore oil platforms around, a submarine has to wend its way safely to its destination. It is somewhat akin to instrumentation flying by aircraft with one exception – there are no ground radars to lock onto. She also has to know, at any given moment, her position with respect to the ground she is moving over with some accuracy. Older conventional submarines periodically updated their positions every time they came to Periscope Depth to charge batteries, with the help of the same aids used by ships. They could also update their positions at any time while they snorkeled and transited at Periscope Depth. Below Periscope Depth, they used the 'Dead Reckoning' method, which noted the compass and Log readings and translated them to a position on the chart. The vulnerability at Periscope Depth to anti-submarine forces resulted in newer conventional submarines staying deep for longer durations, using Air Independent Propulsion systems (AIP). The birth of the nuclear propelled submarine eliminated the need to come to Periscope Depth altogether. The Dead Reckoning method was not accurate or good enough for underwater navigation for long periods, and errors in positional accuracy accrued. Other, more accurate options like *Ship's Inertial Navigation System* (SINS) have since been invented and put in place, but along with it, special charts highlighting the bottom configuration of the sea have been created that are very different to the ones used by ships. Some countries have made these charts for the specific areas that their submarines operate in. These charts are not shared with other nations and are 'classified'. Those that do not have these underwater charts risk their submarines operating close to the bottom of the sea, in shallow water operations, or in volcanic regions. A case in point is the nuclear submarine *USS San Francisco* (SSN – acronym for Submersible Ship [Nuclear] also called the Attack submarine) ramming, at flank (maximum) speed, a submerged mountain in the East Mariana Basin approximately 350 miles south of Guam, on 08 February 2005. She was on her way to Australia at a depth of 525 feet when the 'grounding' occurred. The chart in use at that time did not show the submerged mountain. Fortunately, she survived and returned home. There are many more examples available of underwater collisions and groundings due to inaccuracies in underwater navigation.

Not all portions of the undersea are problematic. Submarines can exploit some areas advantageously. Flat, sandy, sea bottoms are used by conventional submarines to 'sit' on – a process known as 'bottoming'. This helps them to be quieter and conserve their battery power while on patrol in their mission area. A submarine hovering against the vertical face of an underwater cliff as a background, can confuse any ship in the vicinity looking for her as the two are extremely difficult to differentiate even with 'active' sonar. Sinking into an underwater topographically present basin while opposing ships are moving above, shields a submarine from being detected – a tactic often used by some. Hiding behind underwater hillocks or pinnacles again deceives a hunter. As already stated, Deep Sound Channels can be used for long-range underwater communications. Avoidance of surface *ducts* and exploitation of Shadow Zones by moving into them can assist a submarine to get closer to her target without being detected at the usual detection ranges. On the other hand, moving momentarily into a surface duct early, can give her extended ranges of detection of ships. Hiding under density layers screens the submarine from the ship, and remaining on the opposite side of similar layers gives screening protection against opposing submarines.

It can thus be seen that by virtue of being able to move in the vertical plane, a submarine can sometimes favorably exploit the undersea to achieve her missions. The business of undersea operations is indeed complicated, and the environment in the area of operations must be well understood to have that important 'edge' so vitally required to put one across the adversary.

USS SAN FRANCISCO DAMAGED AFTER COLLIDING WITH A SEA BED PINNACLE

THE SUBMARINE

"There are times, when silence has the loudest voice"

- Anon

Like high-end motor vehicles that despite fulfilling an essential need of transportation, are each designed for niche purposes, submarines too can be classified into different categories depending on the roles they perform, even while carrying out the same essential function of undersea warfare.

Clandestine Operations

Consider that there is a requirement to covertly pierce the defenses of an enemy's harbor to destroy a vital target inside. Most harbors are shallow, and anything approaching on surface can easily be detected. A submarine that can go as close to the harbor as possible without being detected is required, so that frogmen or unmanned underwater vehicles can do the last lap and carry out the task. The last lap should be a short distance that can convey frogmen to the targets successfully, or envelope the control range of robotic vessels (more on this in a later chapter) to accomplish it. The submarine, perforce, has to be small. She cannot therefore possibly have a large range of operation because her size will restrict the amount of fuel and battery power she can have. Such a submarine – the midget – is designed and built to go where larger submarines cannot. It follows that when the target is at some distance, the midget has to be towed or carried 'piggy back' and released at a safe distance from the target so that she can stealthily do the last part of the journey by herself. She also has to come back after recovering the frogmen or unmanned vessel, and be towed back. Of course, to get the maximum out of her, she can be given a few torpedoes for self-defense, or mines in lieu, to do that little something more. A pair of torpedo tubes may be fitted, with no scope for reloads due to lack of space. These can be used for attack or self-defense. A few mines in lieu of the torpedoes in the tubes will not make an effective minefield and will be only of 'nuisance' value. Nevertheless, the discovery of just one mine by the adversary can tie down a very substantial part of his harbor defense and minesweeping/hunting forces for days on end. These midgets cannot be deployed for wide-open ocean missions. They are built for *specific* roles and tasks, the example just cited being only one of them..

Shallow Water Operations

If the waters of interest have, say, a long continental shelf gently sloping and deepening seawards as the sea bottom (wholly or even partially), then the submarine should be designed for operating freely and safely in shallow waters. The term 'shallow' here should not be mistaken for the shallow depths to which midget submarines go. We are talking about a submarine that can operate in waters above a continental shelf, seaward of the 20 to 30 meters sounding line, and further out into deeper waters. It need not be capable of diving too deep as its prime requirement is to operate in shallow waters (such a decision about reducing its diving depth can result in significant cost savings).

While the maximum diving depth can be somewhat relaxed, it must have the required endurance to get to its deployment area, carry out its mission, and return. It will be larger than the midget, but definitely smaller than an ocean-going boat. As an analogy, put a large fish and a small fish in an aquarium and slowly lower the water level. At the level where the big fish begins to get uncomfortable, the small one is still comfortable and swimming freely.

Such a submarine, designed for *shallow water operations*, must have the required maneuverability, and be *far quieter* than other submarines. In shallow waters, reverberations enhance the noise of the boat and the sea. This has to be countered. The sonar sensors have to be able to perform through the shallow water noises in the background. The running and search patterns of torpedoes must cater for shallow water operations. Meeting all these special requirements costs! The extra cost must be incurred to make it the specialized platform it is required to be.

Operations in Choke Points

Countries located in enclosed seas with narrow straits as entrance and exit points for shipping, may want to design a short-ranged conventional submarine that can *lie in wait* at these *choke points* for its target. In such a case, the transit requirement may be small and the requirement to stay in its patrol area for extended periods - in the vicinity of the choke point – the primary need. She should be difficult to detect, and that means she must be a very quiet boat, as her area of operations will be small, and her pattern of operations repetitive.

Such a submarine may need to carry less of the fuel that is required for long transits, and have increased battery capacity to remain underwater for long periods in the patrol area. At the same time, she may require a battery charging system (Diesel generator) that will 'top up' the drained batteries in the quickest possible time. She will require a very sophisticated sonar that can differentiate between the myriads of ships, submarines, ferries, fishing vessels, etc that will be moving through the choke point round the clock – again a specialized need.

The choke point may have shallow or deep waters, which is another consideration to be made when drawing up the Staff Requirements for this particular submarine. The submarine that is finally built to meet these requirements would be most suitable for choke point deployments in restricted seas.

Open Sea Operations

A country with a long coastline, or an area of operations far removed, may need a submarine with good endurance and a discreet 'transit speed' that is fairly high, in order to be deployed on offensive missions at some distance away from her base port. Such a submarine must also have 'staying power' in the form of reliability and redundancy in its equipment and systems. She must have an acceptable 'indiscretion rate' (will be discussed later). She should be able to receive messages at great ranges, underwater. She should be able to carry all the armament she requires for long deployments without having to frequently return for reloads. She should be an *ocean going submarine*.

Under-Ice Operations

Submarines *operating under the ice* in the Arctic Ocean have a different set of requirements to those operating in Temperate, Tropical, or Equatorial latitudes (who will require air-conditioning systems). Because of the low temperatures, a different set of oils and lubricants may be required for equipment and machinery. A steam or heating system may be required to 'de-freeze' moveable parts that freeze up in extreme cold conditions. A different set of optical and other instruments and sensors may be required to operate under the ice. The fin/sail may have to be hardened to enable it to pierce through three to four feet thick 'polynyas'. Additional navigational facilities for the Polar region may be required. All these are, once again, specialized requirements.

Anti-Submarine Operations

It has always been said that the best *anti-submarine warfare platform* is another submarine. It can nullify the environmental factors used by submarines to hide, and move with equal advantage in the third dimension. Thus was born the Hunter Killer diesel electric submarine, or the SSK for short. Her role demands all the capabilities required to outwit and outperform her underwater opponent – better maneuverability, better agility, longer detection and weapon range, so on and so forth. To give her this advantage, it may be necessary to design her with reduced reserve buoyancy, give her a tighter turning circle, a faster acceleration rate, and other such features. The SSK is a specialized submarine built with all the required advantages over the other boats. However, looking for another submarine in the wide ocean is like looking for a needle in a haystack. To be able to do this she must be deployed in a properly defined *very high submarine probability area.* Since a 'submarine versus submarine' situation is an extremely rare encounter between conventional submarines, the SSKs could also be given a subsidiary anti-shipping capability with a medium range of operations.

Other Submarine Operations

Then there is the submarine specially built to act as a '*practice target submarine*' for others to fire at as part of their work up to hone their skills. She has to be well padded to carry out her role, and be able to survive the odd direct hits. Practice torpedoes are normally fired with a throw off, or programmed to pass under or over the target. She should have high 'reserve' buoyancy and good watertight integrity. An example would be the Russian '*Bravo*' class submarine built for this very specific purpose. There are submarines built for *submarine rescue work* that carry a DSRV (Deep Submergence Rescue Vehicle/vessel) piggyback, to link up with a boat in distress. They would have all the facilities required to give immediate support to a sunken submarine till help of the required magnitude arrives.

The Staff Requirements for a submarine design evolve from the role it will be required to perform in defined geographical environs, against the type of opposition it is likely to encounter during its life cycle. Since the end of the Cold War, more and more small to mid-size navies have been placing orders for diesel-electric submarines to meet their protective requirements. A country without any pretensions of being a super power will have roles influenced by its geographical location, and that of its likely adversaries. Some of these countries may be part of some Alliance, in which case they are likely to be asked to deploy their submarines in the same role that they have originally been designed for, in support of that Alliance.

Having drawn attention to the different requirements, and therefore to some of the different types of conventional submarines that are required to be built, and have been built, we shall now look at the nuclear club and their different requirements resulting in different types of submarines that have come up in their sphere, as part of the underwater scene.

The Nuclear Submarine

A nuclear submarine is *not nuclear* propelled. It is steam driven. This makes it noisier than conventional submarines in certain frequency bands. The prime mover is noisier. Consequently, tremendous efforts in terms of costs and technology go into making the boat as quiet as possible. Turbines are turned by steam that, in turn, turns the propeller shaft through gearboxes. The shaft consequently turns the propeller. The nuclear plant manufactures the steam required by the steam turbines.

Nuclear submarines may be deemed to be *true* submarines. They can remain and operate dived for months on end, and move about at sustained high speeds. Electrolysis of sea water provides constant supply of much needed oxygen for breathing. The limiting factor is human endurance (crew fatigue) and the amount of victuals that can be embarked. Different approaches have been made to exploit the boat's capability to remain at sea for long periods. A two-crew system per boat has been devised by the US navy and first tried out on their '*George Washington*' class boats in the 1960s, with one crew handing over to the other during a brief visit to harbor for the switchover and replenishment of victuals. (This could also be affected in a forward base to obviate the need for the boat to come all the way back to her homeport. The crews then have to fly out and back).

In another drive, every attention has been given to enhancing crew habitability conditions with strict environmental control and facilities like a gymnasium, sauna, jogging track etc thrown in, so that they could stay out at sea 'a little longer,' in more pleasing surroundings. Despite all this, the fact remains that the machine – the nuclear submarine - outlasts the man who mans it at sea. These submarines are generally deployed in waters deeper than 30 fathoms as they are not too comfortable in shallower waters. This is not to state that they cannot make forays of short durations into shallower waters for specific operations.

Like the diesel-electric submarine, nuclear submarines are also built with specific abilities, for specific roles. The Cold War, the TRIAD requirements (capability to carry out a nuclear strike from land, the air, and from sea), and the hunt for a second strike capability as a deterrence, saw both sides build large nuclear submarines to house missiles aimed at each other's countries and important installations. These giant behemoths are referred to as *SSBN*s – the acronym for Ship Submersible Ballistic Nuclear. Some also refer to them as "Strategic Ballistic Missile Submarine, Nuclear Powered". These are meant exclusively for strategic roles, but they generally have a few torpedo tubes for self-defense. The first generation Soviet SSBNs were the '*Hotel II*' class with their surface-launched SS-N-4 missiles that made their appearance in 1958. The missiles were soon replaced by the underwater launched SS-N-5s. They were noisy boats even by standards of the fifties and could easily be located and tracked at sea by the Americans.

The first American SSBNs were the '*George Washington*' class that carried 16 Polaris A3 missiles (range 3500 miles) and made their appearance in 1959. Only five were built. Over the years, successive, improved, more sophisticated, classes were built by both sides with progressive increase in missile ranges. Multiple warheads (MRVs – Multiple Re-entry Vehicles, and MIRVs – Multiple Independently targetable Re-entry Vehicles) replaced single warheads. Their deployment patterns also changed. Post-Cold War, we have only one superpower wielding its influence worldwide. At the time of writing this book, the U.S. has 18 *Ohio*-class submarines, of which 14 are Trident II SSBNs, each capable of carrying 24 SLBMs and providing strategic sea force levels to meet their requirements. (The first four which were all equipped with the older Trident I missiles have been converted to SSGN's each capable of carrying Tomahawk guided missiles and have been further equipped to support Special Operations).

Russia inherited most of the forces of the former Soviet Union and struggled to maintain them even as the economy, during the changeover from the old to the new system, took a beating. Russia has been able to maintain an active force level of six SSBNs of the '*Typhoon*' class, and they presently provide her with the strategic sea-borne strike effort and deterrence she is seeking. The '*Typhoon*' is the largest submarine ever built to date, with a displacement of 26,500 tons, dived. The twenty SS-N-20 missiles onboard (since replaced by more accurate ones) have a range of over 4500 miles. Each missile has 10 MIRV heads. She can hit many cities in the United States from her homeport in the Kola Peninsula, *without having to sail out*: the only reason she would do that is to avoid missiles aimed at her. The '*Typhoon*' Class is already on its way out and being replaced by the new '*Borei*' Class SSBNs (with SS-N-28 missiles) with third generation technology that promises to make them extremely quiet and difficult to detect or track. Four of them have been ordered, with another four approved, fourth generation submarines (with SS-NX-30 missiles) in the pipeline. The '*Borei*' Class submarines have a lesser displacement than the '*Typhoon*' Class but provide the necessary punch.

These new generation submarines on both sides are extremely quiet and sophisticated nuclear propelled submarines. To make them quieter, the most advanced technology available has been used, and the cost is mind-boggling. Both countries additionally continue to operate their older SSBNs.

Apart from these two nations, we have the United Kingdom, France, China, and India also operating strategic submarines of the SSBN or SSGN variety. Their force levels, however, are nowhere near that of the two major powers just described, but are considered adequate to meet their own legitimate requirements. However, when operating with allies or with coalition forces, they would be forces to be reckoned with.

The story does not end there. Newer and better platforms are regularly emerging from drawing boards and will put out to sea as replacements. Such gigantic, behemoths cannot be allowed to operate at sea unchecked. Both NATO and WARSAW pact countries endeavored to *tail their opponent's SSBN with a submarine* that had the agility and ability to do so. Thus was born the nuclear powered Fast Attack Submarine. She initially carried no missiles, had torpedoes as her only weapon, and had the ability to move fast, move quietly, and follow the SSBN wherever she went. With the first indication of the balloon going up, even before the latter could launch her destructive missiles, the Fast Attack Submarine would attack and destroy the SSBN being tracked – or so it was planned. This Fast Attack Submarine got to be known by the acronym **SSN**. She was, and is, versatile enough to be deployed for either *operational or tactical roles.* The SSBN was now vulnerable, and could not be allowed to proceed to sea unprotected.

The SSN got her next role – to escort the SSBN, detect the opposing SSN approaching to tail her or tailing her, distract her, and let her own SSBN slip away and get 'lost' in the wide open ocean, to be rejoined later. This required a series of deft and bold maneuvers underwater, backed by very sophisticated detection and trailing abilities. During the Cold War some of these ended up with underwater collisions that had been closely guarded as secrets by both sides. Details of some of these collisions have been de-classified and are now available

The SSN showed herself to be a very agile and versatile boat. She began to be used for more and more different roles. She could be assigned any one or a combination of the following tasks: -

- Tailing an opposing SSBN,
- Hunting for opposing SSNs,
- Escorting own SSBNs,
- Hunting down diesel-electric submarines,
- Escorting and shielding a Carrier Battle Group from enemy SSNs (role presently undertaken by American SSNs only),
- Interdiction of a Surface Action Group (SAG) (in war),
- Attack a convoy, an Amphibious Task Group, or a large formation (in war),
- Carry out clandestine and intelligence gathering missions.

In recent years, with the introduction of Submarine Launched Cruise Missiles (SLCM) onboard SSNs, the tasks of attacking land targets and enemy airfields have also been added to the listed roles.

An SLCM, like the 'Tomahawk', is a *grey area* missile in that it can have either a conventional warhead or a nuclear tip. The Russian *'Victor III', 'Sierra I&II', 'Akula'* and *'Yassen'* Classes, the US *'Los Angeles', 'Seawolf',* and *'Virginia'* Classes, the British *'Trafalgar'* and *'Astute'* Classes, the French *'Barracuda'* and *'Rubis'* Classes, and the Chinese *'Shang', 'Han'* and *'Type 095'* Classes, fall in this category, to mention just a few.

Like the conventional Russian *'Bravo'* class, the Americans had the *'Benjamin Franklin'* class of SSNs that were specifically deployed only for *delivering special operations forces.* There was one more type of nuclear boat and that was the one that carried non-ballistic, cruise missiles, and known by the acronym *SSGN.* The Soviet navy had the *'Charlie II',* the *'Echo II',* the *'Oscar'* class, and the *'Papa'* class (only one built). They also had a diesel electric submarine with this capability, namely the *'Juliet'* class built in the early 1960s. These were submarines built to specifically target US warships, and their aircraft carriers in particular. As mentioned earlier, four of the original US *'Ohio'* Class SSBNs have been converted and are now operating as SSGNs for "from the seas" operations. They can carry up to 66 Special Operation Forces to be released close to enemy coast. They also can carry a platoon of Navy SEALS who are released through two 'Lock Out' chambers with access to sea delivery vehicle submersibles that can operate in '6 hours out and 6 hours back' mode. This is apart from the up to 154 Tomahawk missiles that they can be armed with. The Chinese and Indian Navies also possess SSGNs in their inventory with both ballistic as well as cruise missiles.

The characteristics of a particular type or class of submarine identify her technical capabilities, and her ability to give off her maximum during war. An understanding of these characteristics is required to allot her realistic missions during when she can use her weapons and technical capabilities optimally. The characteristics that require attention are: -

- *Displacement* – on surface and when dived.
- *Speed* – maximum and economical on surface and when dived during normal operations. Maximum and minimum silent speeds in a tactical environment.
- *Operating range* with normal loading of fuel and with extra fuel (carried in ballast tanks) – applicable and relevant in respect of conventional boats.

- *Number of propeller shafts.* Most submarines now run on a single shaft with one propeller. Some have thrusters that can be used skillfully in a tactical environment. The single propellers are getting more and more sophisticated with varying numbers of blades and skewing to make them as silent as possible, even at high speeds.
- *Reserve Buoyancy.*
- *Energy producing ability* (relevant to conventional submarines. Number of diesel generators that can be coupled on to the motors simultaneously, or separately. Power of diesel generators).
- *The type of batteries* and battery capacity
- *Dimensions of the submarine* – Length, Breadth, and diameter of the pressure hull.
- *Diving Depth* – (working depth, maximum operating depth, and designed depth).
- *Weapon package.*
- *Endurance.*
- *Sensors carried* – (sonars, electronic, radio etc).
- *Crew strength.*
- *Strategic and tactical capabilities.*
- *Quietness.*

Most submarines are now propelled on a single shaft with one propeller or 'pump-jet propulsor'. Some have additional thrusters that can be used skillfully in a tactical environment. The single propellers are getting more and more sophisticated with varying numbers of blades and skewing them to make them as silent as possible even at high speeds is an on-going process.

It can thus be seen that, just as the word 'motor vehicles' has different connotations, the word 'submarine' also needs to be looked at for further classifications. Very few nations have design bureaus and a submarine building capability. These have the luxury of building boats to meet their *exact requirements.* Logically, economic considerations dictate that after protecting their own interests, an *export version,* with de-classified equipment be offered on sale in the international market. Some do this, as there are enough takers for this relatively cheaper, but potent, platform. There is a warning here. Purchasing a boat designed for a particular role and then exploiting her in a very different role would be unadvisable. The preferred option is to ask those with a building capability to design and build a boat to meet *your* specific requirements. Of course, the best option is to *build your own boat to meet your own requirements.*

Without doubt, a submarine is an attacking platform. She has the advantage of freedom of movement in the vertical plane in water. She has better knowledge of the water conditions around her and in her area of deployment than surface ships or maritime aircraft. She is quiet and difficult to detect, track, or tail. She can detect ships much before they even get an inkling of her presence. She is expected to use these attributes to advantage in offensive operations whenever deployed, in peace and during hostilities. In peacetime deployments she is expected to move clandestinely into sensitive areas and gather information that will be useful during hostilities. Gathering useful information about sea conditions, submarine probable areas, harbors and coastlines, other ships, their common routes, traffic densities in particular areas of interest, characteristics of warships, their habits and pattern of operations, and other such information, is part of her every day functions at sea.

Pre-deployments for mine laying or offensive operations are natural patterns of operations for her just before hostilities. During hostilities, she must be the main concern for worry for the adversary. She must seek, hide, stalk, harass, and destroy. She must be the epitome of a potent offensive threat.

Many books have been written on the evolution of submarine designs, and the principles of handling a boat underwater. The transition from a hull design akin to that of ships and suited for high-speed transits on surface (but with slow underwater speeds) to one that was more suitable for higher underwater speeds: the 'albacore' or 'tear-drop' hull that attracted some: the subsequent discovery that long narrow hulls offer less drag than the tear-drop hull, have all been commented upon extensively. Design improvements on propellers, the number of blades desirable, the extent of skewing, advantages of pump jet propulsors, the desirable position of the rudder *vis-à-vis* the propeller, the cruciform rudders versus the vertical rudder, the various positions the hydroplanes have, their advantages and disadvantages etc., have also been covered extensively in literature already.

VICTOR III CLASS SSN

OHIO CLASS SSBN

SUBMARINE CONTROL

"Of all the branches of men in the forces there is none which shows more devotion and faces grimmer perils than the submariners."
Sir Winston Churchill

Like a feather floats weightlessly in the air, so does a submarine move about with neutral buoyancy below the sea. When she is needed to come to the surface, she is made positively buoyant. Conversely, if needed to dive, she is given negative buoyancy (made heavier). Once underwater, she adjusts her weight and brings it from negative to neutral or zero buoyancy to maintain the desired depth.

The submarine does these maneuvers with the help of *Ballast Tanks* that are situated between her outer hull and the pressure hull. If all the water is expelled from the ballast tanks and replaced with air, the boat remains positively buoyant and stays on the surface. When the ballast tanks are flooded with water again, she begins to sink, gaining momentum as she goes down, till the negative part of her buoyancy is removed. With neutral buoyancy, the submarine maintains depth.

Similarly, when water in the ballast tanks is replaced with air again, she gains positive buoyancy and breaks surface. Double-hull submarines have two hulls: the outer hull or the 'casing', and the inner hull or the 'pressure hull'. In the space between the outer and inner hulls, apart from the Ballast Tanks, she could also have her sonar domes, the drives for her hydroplanes, high-pressure compressed air bottles, a portion of the tubes of her torpedoes or missiles passing through, her snorkeling system and ventilation system trunkings, and a host of piping and other equipment. This area is 'free flooding' meaning that sea water fills the space once the submarine dives. Single hull submarines have an inner pressure hull and a combination of ballast tanks and an enveloping casing inside which the other equipment mentioned above, run. Some large submarines have more than one pressure hull enveloped by one overall casing, with direct watertight openings between pressure hulls.

Inside the submarine she has two or more *Compensating Tanks* located near her centre of gravity and centre of buoyancy to take in or let out water equivalent to the amount required to be moved in or out to maintain neutral buoyancy. At the forward end of the submarine and at the aft end she has *Trim tanks* that move water between them – forward to aft: aft to forward – to compensate for tilts that may develop due to various reasons.

To propel in the water she has one or two *propellers*. Some modern boats have a *pump jet propulsor* in lieu. A *rudder* permits her to change course as required. To change depth underwater, she has a set of *forward hydroplanes* that are effective only so long as she is moving in the water. These could be located either in the forward portion of the submarine outside the pressure hull, or up on the fin/sail. To control the fore and aft tilt or maintain an even keel while underwater, she has a set of *aft hydroplanes* that are also effective only when the submarine is moving. While propelling, the submarine can make use of the forward and aft hydroplanes to compensate for minor deviations from neutral buoyancy, as also to change depth and control the angle of tilt. On some submarines, the forward hydroplanes are substituted with planes on the fin or sail.

As she goes deeper, the submarine's hull gets compressed by the increasing external pressure of water column being exerted on her. She then displaces less water progressively with compression. As she displaces lesser and lesser water, she becomes heavier and heavier, and sinks. To get her back to neutral buoyancy, she has to be made lighter by that amount of water that has made her heavier, using her Compensating Tanks (expelling water from them).

Sea water is not a homogenous medium. The temperature, salinity, and density, vary constantly, and directly affect her buoyancy. With every change in sea water conditions the submarine will tend to move from neutral to positive or negative buoyancy. The more saline or denser the water, the more positive will be her buoyancy. At the entrance to a river flowing into the sea, the water will be less saline. Consequently, the submarine will start sinking because in those waters she will be more negatively buoyant. Continuous adjustments to weight have to be made with the help of Compensating Tanks when the submarine is underwater, to maintain neutral buoyancy.

If, during the course of days at sea, the dirty water tanks or the slop drain tanks get filled up as a result of the crew using up fresh water for personal hygiene, those tanks will have to be pumped out overboard and got readied for more intakes. As the tanks are pumped out, the boat starts getting lighter and coming shallow. To get back to the original depth, a certain amount of weight equal to the amount of water being pumped out has to be put back into the submarine. The Compensating Tanks are flooded with sea water for this.

Let us now look at another factor; movement in the water as a result of propulsion. When a neutrally buoyant submarine starts to propel in the water, the forward movement makes her bow lighter and rise upwards. Her forward and aft planes can be brought into play to check this tendency to rise to the surface. Some water from the Aft Trim Tanks may be transferred to the Forward Trim Tanks to counter the effect. Alternatively, some weight can be added through the Compensating Tanks to her present state, to counter this urge to rise. A combination of all three actions may be taken too, to counter this effect.

While underwater, and neutrally buoyant at a particular depth, if the crew is called for lunch or dinner or supper, there will be movement from various compartments towards the galley to help serve themselves a meal. This movement will affect the even plane of the submarine which is at zero buoyancy. This has to be adjusted for through Trim tanks. When they move back to their respective positions thereafter, a counter effect will occur that will have to be corrected again. If they wash up before they move to the galley, then a certain amount of water moves from the water tank to the wash basin, and from the wash basin into the slop drain tank. That amount of water has been physically displaced from one tank into another, and this, in a neutrally buoyant boat, will cause another tilt that will have to be compensated. The *Trim Tanks* are utilized to transfer sea water forward to aft or aft to forward to compensate for these types of tilt.

In a dived submarine, moving up or down (changing depths) and across varying densities of water, both the Compensating and the Trim tanks are frequently in use. This is done in modern submarines through automated systems.

A submarine has what are termed *as Fixed Weights* and *Variable Weights*. Fixed weights are permanent fixtures onboard a submarine whose locations and weights do not normally change – machinery, systems, compressed-air bottles, auxiliaries, etc. Then there are moveable weights that change every time a submarine goes out to sea – number of torpedoes and missiles, number of personnel, their luggage, victuals, fuel, oils and lubricants, distilled water, fresh water, stores, spares, and so on. Their disposition inside the submarine may also vary each time the submarine goes to sea. While the disposition and weight of fixed weights are compensated for on a more or less permanent basis, the disposition and weights of moveable objects changes every time a submarine puts out to sea.

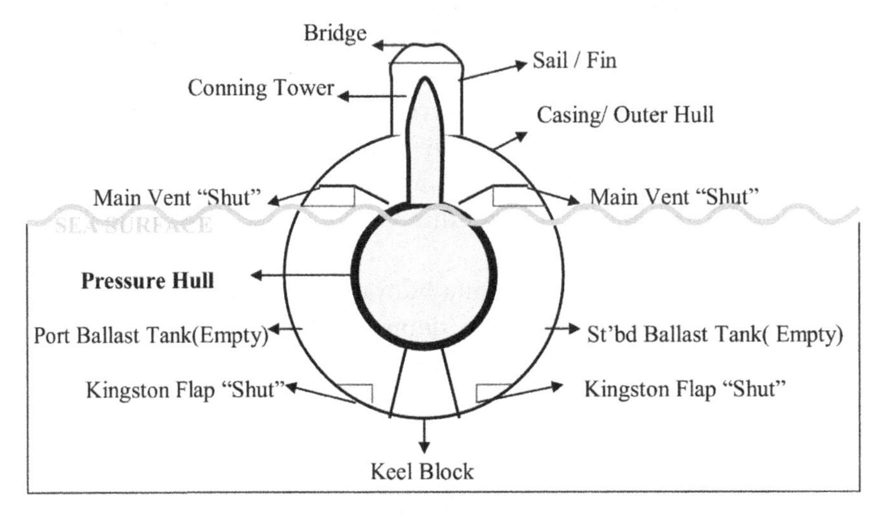

Fig 1 – SUBMARINE ON SURFACE

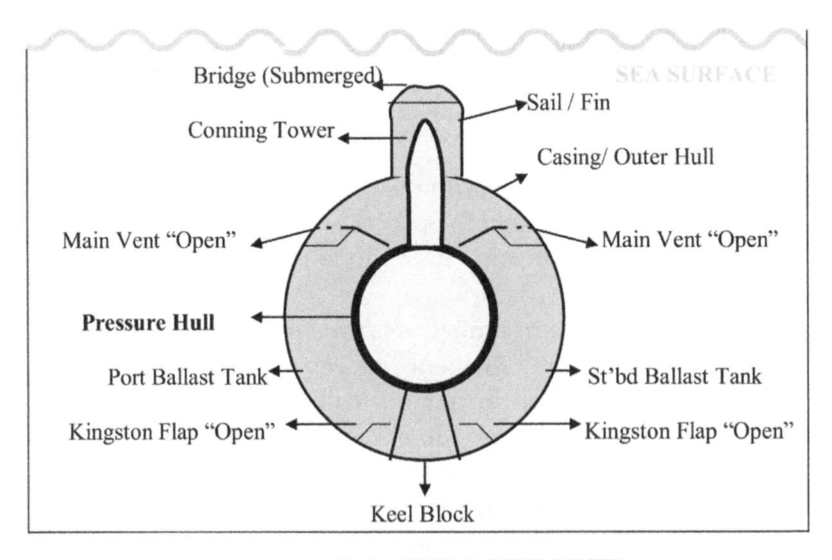

Fig 2 – SUBMARINE DIVED

The Engineer Officer has to work out the Trim of the submarine after taking stock of all moveable weights and their location inside the hull. This is a theoretical, but meticulous, exercise carried out in harbor taking every little variable weight into consideration. As part of these calculations, he finally compensates for trim and overall weight by practically adjusting water in the Trim Tanks and filling the Compensating Tanks with a certain amount of water. This has to be confirmed as correct at sea and accurately re-adjusted so that when dived, the submarine can be quickly brought to neutral buoyancy, and can be handled by the crew with the required agility. If his calculations are accurate, he may have to carry out very little adjustments to the amounts of water he has put in the Trim Tanks and Compensating Tanks. How accurate his calculations are is determined by a practical dive at sea known as the *Trim Dive.*

A *Trim Dive* is therefore initially carried out by the *crew every time a submarine puts out to sea.* It is also the first dive she carries out after completing a refit. It is done as soon as the submarine leaves harbor and gets to deep enough waters to carry out the dive. It is a cautious dive that is carried out slowly and in stages. The submarine is propelled on surface to the area where the Trim Dive is to be carried out. All personnel onboard close up at their work stations. She is then stopped and prepared to dive. The Ballast Tanks are flooded in stages. Ballast Tanks are sub-classified as Forward Group, Center Group, and Aft Group. The end groups of Ballast Tanks are first flooded to get the submarine down to what is referred to as the '*Trimmed-Down Position*'. The submarine still has positive buoyancy but only marginally so. Keen eyes in the Control Room watch the equivalent of a spirit level bubble and the depth gauge to see how the submarine is behaving. If calculations are wrong and she is too heavy, she will show indications even before the center group of Ballast Tanks are flooded.

Clever engineer officers intentionally keep the boat a little light by taking in less water in the Compensating Tanks then their theoretical calculations suggest. During the Trim Dive they then take in water in stages to get the right amount in. If the disposition of water in the Trim Tanks is not accurate, she will settle into a bow down or stern down attitude. Some adjustments to the water in the Trim and Compensating Tanks may be made at this stage if required before proceeding further. It is good submarine practice to get the control surfaces of the Forward and Aft Planes into play at this stage for better control, and so the order to propel is given, and as the boat makes way the Center Group of Ballast Tanks are flooded.

The submarine now begins to go ahead and down with the Forward Planes controlling the change in depth and the Aft Planes controlling the angle of dive (the tilt or 'bubble'). The order is given to get down to Periscope Depth. The Search Periscope and some of the other retractable masts (Radar, communications, etc) are raised and vigilance kept of the above water scene. More adjustments are made to the quantity of water in the Trim Tanks and Compensating Tanks till the boat 'catches a good Trim'. She is then taken to *Safe Depth* after all retractable masts have been lowered back into the sail plane or fin. *Safe Depth* is that depth at which the top of her sail or fin is clear of the deepest draught vessels known or expected to be plying through that area. The *Safe Depth* is different for different classes of boats. Here again her Trim is adjusted. The submarine may be taken down deeper, but generally it is acknowledged that she has caught a good trim if she settles well at *Safe Depth.* The Trim Dive successfully completed, the submarine can proceed in accordance with previous orders, to carry out her mission.

In contrast to a Trim Dive, a normal dive is the type of dive a submarine carries out at sea whenever she has come up to the surface or Periscope Depth, and has to dive again. A *Normal Dive* is executed by the crew when the boat is well trimmed. All the actions are similar but take less time as they are done in quicker succession, smoothly.

An *Urgent Dive or Crash Dive* is carried out by submarines as a part of evasion in order not to get caught by an undesirable in the vicinity. It can be executed from the surface or even from Periscope Depth. Those submarines that have large reserves of buoyancy have what is termed as the *Quick Diving Tank or the 'Q' Tank* that is filled in a hurry to give them that extra negative buoyancy that is needed to overcome inertia and get down fast. This tank is flooded in addition to the other ballast tanks that are flooded, to go down. However, this is a tricky tank and needs to be fully blown empty once the submarine begins to gain momentum while going down as the negative buoyancy it creates is considerable. If not blown on time, the submarine becomes very difficult to control and settle at the desired depth.

All submarines have *a maximum operating depth, a maximum diving depth, and a crushing depth.* These are arrived at after superimposing a safety factor as reserve. That was why one heard of submarines during the Second World War going beyond their 'crushing depth' and returning without getting crushed. The safety margin saved them. The first of class is taken down to the maximum diving depth during acceptance trials to prove her worth and design. All other submarines operate only up to the maximum operating depth, which is less than the maximum diving depth. This is because metal fatigue sets in every time a submarine's hull is subjected to extreme compressions and decompressions leading to lessening the life of the pressure hull. The safety factor built into these depths permit the commanding officer to feel free to maneuver the submarine between the surface and her maximum operating depth as often as he wishes, during her operational cycle. Some navies impose further restrictions on the number of times a submarine can go down to her maximum operating depth during an operational cycle in peace-time as an additional precaution.

During an extensive refit of a submarine, a number of pieces of machinery are removed for refurbishment or servicing. Some modifications and upgrading may also be carried out. Some items may be re-sited and fresh ones sited. In effect, a change in the dispersion and weights of moveable weights may take place. These are adjusted for by rearranging weights in the keel block, and recorded in the engineer officer's Trim Book, as they affect trim. Variable weights, as always, also undergo changes in locations and weights. Therefore, after a refit, it is mandatory for a submarine to proceed to sea with minimum variable weights onboard to carry out a Trim Dive followed by a *Deep Dive* before she embarks her weapon outfit and other variable weights. The Deep Dive is, again, a cautious dive in stages, all the way down to her maximum operating depth. The effectiveness of the refit and the correctness of the engineer officer's Trim Calculations are once again checked out. It is normally recommended that a Deep Dive be carried out at least once in every operational cycle of the boat.

Just as adjustments to *Trim* and *buoyancy* are continuous in a dived submarine, so are movements of the *rudder* and the *hydroplanes.* All these may be handled manually or through automated systems from the Control Room. The control surfaces of the rudder and planes, and variations in submarine speed that affect them, all play a part in controlling the boat. One may liken submarine control to controlling an aircraft, with one big, unique, difference – the aircraft is handled by one, maximum two personnel sitting side by side. The submarine is controlled by far more personnel, in different compartments, and calls for greater co-ordination and teamwork. It also calls for imposing trust in each other's capabilities and performances implicitly. To reach this level, extensive training and work-up as a crew is carried out – far more than in surface ships.

DEPLOYMENT STRATEGIES

"Of all the types of naval warfare that are automatically conducted at the operational or the strategic level, it is the submarine offensive and the resultant anti-submarine campaign that is the most important and least understood."

R.Adm. Raja Menon

Gaining three dimensional control of the naval battle space is a prerequisite for joint power projection from the sea. Control in the air, on the sea, and under the sea is a must. However, to make power projection effective and sustainable, it is necessary to have control of the sea. That control can sometimes be exercised even by submarines alone.

The ability to move about underwater stealthily and undetected gives the submarine many tactical advantages over surface and air elements. Well before the outbreak of hostilities, she can be so deployed as to use the water space in the vicinity of an unsuspecting hostile force (or its harbor), track and observe it, and then alert own forces in advance of the enemy's movements.

It may be argued that this becomes redundant when own forces have the luxury of obtaining satellite pictures to track movements, but many nations having this facility still opt for deploying submarines and positioning them in advance, so as to be able to physically act within minutes of the 'balloon' going up. She can be closest to the scene and therefore one of the first to be called upon to act. She could be deployed to lay mines in strategic areas that can be activated at a later date to create havoc on any force passing through the area, or, alternatively, stealthily gather vital intelligence from areas where ships and aircraft are uncomfortable to go.

The attributes, traits, and characteristics of this underwater platform lend themselves to evolving deployment strategies to meet different requirements in different situations and environments.

A World War Chronicle of Submarine Action

The Atlantic Theatre. During World Wars I and II, submarines were deployed by both the 'Axis' and the 'Allies' primarily *for naval blockade*, or *for sinking of both merchant and war ships to paralyze the opponent's military industry* and war effort, or force him to commit and deploy a large number of antisubmarine surface and air units against these underwater menaces that he would have liked to deploy in different roles elsewhere. The submarine was used as a classic platform in the war of attrition where enemy losses were planned to be higher than own losses, both in terms of men as well as ships. To be able to attack at any place, at any time, with total surprise, and inflict heavy losses on the victim, made the submarine a highly profitable weapon. With a crew of less than a hundred, she could do more damage than a surface ship with a crew many times that number. Cost-wise, she was a cheaper option. To counter or neutralize her, a disproportionately large force had to be deployed.

The German navy was quick to realize the potential that this weapon platform possessed and exploited it enthusiastically. During both the World Wars, they used submarines as their main weapons at sea. However, between the two Wars, the advantages of developments in the field of technology was not noticed and seized by her or any of the other major powers. Thus it was that although submarines were the only platforms that came nearest to being decisive during World War I, they were neglected in the immediate years that followed. During the World War II, these same platforms with very little improvements in design and weaponry gained initial successes, largely because of a change in deployment tactics. Wolf pack tactics were utilized whereby they deployed in a spread out fashion in the North Atlantic Ocean, and the one discovering an enemy convoy of ships 'homed' the others on to it. A collective attack followed.

In the absence of sophisticated radars in the initial phases of the War, they moved about on the surface and dived only to close in and launch attacks or to evade anti submarine ships, and surfaced once again after getting away to a safe distance. In the melee and confusion that ensued, they were able to slip away quite easily to fight another day. Most of such attacks were conducted at night. However, as the war progressed, with improved radars appearing on enemy platforms, and more and more aircraft being deployed to hunt and destroy submarines, German submarines were forced to spend more and more hours underwater and less time on surface.

By their very design – the hulls were shaped for high *surface* speeds and low underwater speeds – this restricted their ability to cover large distances in the short time they used to before radar and aircraft were deployed against them. The ability to get to distant patrol areas relatively quickly took a beating. Then the gap between technology and doctrine began to lessen. The subsequent invention and exploitation of the 'schnorkel' countered this drawback to some extent. Other new technical and electronic inventions soon followed and helped the submarines to be more alert and perform better underwater. In the Atlantic Ocean, the Germans exploited submarines better than any of the other nations during both the World Wars. They lost the Battle of the Atlantic because they had insufficient boats (funds asked for constructing submarines were diverted to build aircraft for the Luftwaffe), and the Allies were able to replace sunk ships and build more, in increasing numbers, and in quick time. The Allies were also able to deploy around 25 ships and a 100 aircraft per submarine in the North Atlantic, which put the odds heavily against the German U-boats.

The Pacific Theatre. Japan had, perhaps because of the vast extent of the Pacific Ocean, what was easily the most diverse submarine fleet of any nation in the Second World War. From the smallest to the largest, these included manned torpedoes, midget submarines, medium-range submarines, specialized supply submarines (many for use by the Army), long-range fleet submarines (some of them carried an aircraft each), submarines with high submerged speed, and submarines that could carry multiple bombers. They also had the most successful torpedoes of all countries that possessed submarines. Despite their obvious prowess in submarine building, they achieved very little during the Second World War. This was because their submarines were mainly employed against warships that were fast, maneuverable, and well defended - unlike merchantmen.

The Japanese naval doctrine was built around the concept of fighting a single decisive military battle. This was what was successful at Tsushima some 40 years earlier, and so they saw no reason as to why it would not be successful again. They considered their submarines to be apt in scouting operations, whose main role was to locate, shadow, and attack 'Allied' *naval* task forces. This approach met with some early successes in 1942 when they sank two fleet carriers, one cruiser, and a few destroyers and other warships, and also damaged two battleships, one fleet carrier (twice), and a cruiser. However, as Allied intelligence, anti-submarine technologies, tactics, prowess, and numbers in warships improved, the Japanese submarines were never again able to achieve this type of success.

There are many who argue that the Japanese submarine force would have been better used against merchant ships, patrolling Allied shipping lanes, instead of lurking outside naval bases and moving with their surface warships. The Japanese submarine fleet is credited with sinking only 184 merchant ships totaling 907,000 GRT during the entire Second World War. This figure is far less than what was achieved by the Germans (2,840 ships of 14.3 million GRT), the Americans (1,079 ships of 4.65 million tons), and the British (493 ships of 1.52 million tons). It may be argued that the density of movement of warships and merchantmen were higher in the Atlantic than in the Pacific at any given moment in time, or during peak times. On the other hand, it has also been suggested that unrestricted warfare by Japanese submarines along American sea-lanes of communications, their supply routes in the Pacific Ocean, around the American west coast, the Panama Canal, and the approaches to Hawaii, Australia and India, would have caused the 'Allies' more difficulties than did the warship sinking's that were actually achieved. In hindsight, losing a significant number of merchant ships, and also needing to spread meager defenses even more thinly along two coasts, would surely have had some substantial consequences for the United States in 1942.

Till the Americans were attacked in Pearl Harbor and drawn into the War in December 1941, their strategy for submarines was to sink whatever they could of an enemy's *military and merchant fleet.* At that time, according to American claims, their submarines were plagued with sub-standard and unreliable torpedoes onboard. There were more 'misses' than 'hits'. It took a while to improve them and then the strategy was modified in the Pacific theatre to go for *unrestricted warfare* that included Japanese merchantmen, troop carriers, military ships, and even fishing vessels. In 1944-45 the tonnage lost by the Japanese to American submarines was higher than their capability to replace them. By the end of the war more than 50 percent of all Japanese tonnage was credited to submarines. American, and some British and Dutch submarines, strangled Japan with this *unrestricted warfare policy* by sinking its merchant fleet, intercepting many troop transports, and cutting off nearly all the oil imports essential to weapons production and military operations.

While this may look impressive, it must be stated that the Japanese were nowhere as efficient in escorting their vessels in the Pacific as the Allies were in the Atlantic, and therefore fell as easy prey.

Looking back at the performance of submarines during World War II, one can say that without exception, all nations utilized their submarines on operational – tactical missions. Strategic missions were beyond them because equipment technology and submarine support organizations lagged behind the demands of such missions. If direct and indirect costs are taken into consideration, the German and American submarine campaigns imposed significant logistical constraints on their opponents, and were cost effective. The former delayed victory while the latter hastened it. As it happened, the cumulative effect of operational and tactical victories by American submarines in the Pacific eventually led to strategic gains at a later stage.

The Cold War. Post World War II, and with the introduction of nuclear power propelled boats with their weapons of mass destruction onboard, the Cold War saw deployments of altogether new dimensions. The United States and the Soviet Union fought a vast, undeclared and sometimes chillingly ominous Cold War under the sea. During the early years of the Cold War (mid–1960s to 1970), Soviet submarines ventured into the Atlantic Ocean and operated at extreme ranges of their strategic, nuclear tipped, missiles targeted at the United States. These were the diesel-electric 'Golf' class Ballistic Missile Submarines (SSB), and the nuclear propelled 'Hotel' class SSBNs. Both carried SS-N-5 missiles that had a range of 1400 kilometers. They patrolled at a distance of around 300 kilometers from the coast, between Florida and Maine, and this large area was defined as *a very high submarine probability area.*

With the introduction of the 'Yankee' class SSBNs, equipped with the 3000 kilometers range, liquid fuelled, SS-N-6 missiles in 1968, the patrol areas moved further seaward, and covered a larger area. Two to three 'Yankee' class submarines were on constant patrol in the Atlantic. Each carried 16 missiles, some with MRV heads (Multiple Re-entry Vehicles). By the mid 1980s, the follow on to the 'Yankees' – the 'Delta I' class SSBNs – began to be deployed in the Atlantic Ocean. They took over the area further north to the one covered by the 'Yankees' that fell east of Newfoundland and south of Greenland.

Equipped with SS-N-8 missiles (8 launchers onboard) that had a range of 9100 kilometers, they were able to cover the entire United States from their patrol areas while the 'Yankees' continued to patrol in their old areas.

In the late 1980s, the lengthened versions of the 'Delta I' class – the 'Delta II' and 'Delta III' class submarines – replaced the 'Delta I' class submarine. They withdrew from the Atlantic and began to operate from the Norwegian Sea in an area between Norway, Greenland, and Svalbard. The 'Delta II' carried SS-N-8 missiles while the 'Delta III' had SS-N-18 missiles with MIRV heads (Multiple Independently Target able Re-entry Vehicle) and a range of 6500 kilometers. The 'Yankee' class submarines were redeployed in the Atlantic Ocean off the western seaboard of Europe, against NATO forces. Towards the end of the Cold War, in the late 1980s and 1990s, the 'Delta IV' class and the 'Typhoon' class SSBN took over from the 'Deltas II' and 'Delta III', and with their 8300 kilometers range SS-N-23 and SS-N-20 missiles respectively, were able to operate from just off their own heavily defended harbors north east of the Kola Peninsula, with relative impunity.

The Soviet strategic submarines found themselves being increasingly tailed by American submarines in the open seas. To give protective cover, Soviet SSN submarines accompanied them. In order to get to their destination, they had to run the gauntlet of static seabed sensors (SOSUS) that was spread across the shallow ridges of the ocean between Greenland and Iceland, and between Iceland and the western seaboard of the United Kingdom (GIUK) by NATO. SOSUS was monitored from Norfolk, to keep a track of Soviet submarine movements in and out of the Atlantic Ocean.

The challenge for the Russian SSBNs was in trying to avoid being picked up and to slip away into the wide ocean, and many managed to do that by resorting to varying disguising tactics that included moving under their merchant ships, getting their accompanying SSN submarines to make noise and draw the listeners attention to them, and the likes. More often than not, lying in wait for them on either side of the GIUK cordon would be an SSN of the NATO forces that would then tail the unfortunate as part of the cat and mouse game. As both super-powers increased the ranges of their nuclear-headed strategic missiles, they withdrew their strategic SSBN submarines further and further away from their original patrol areas (but always within strike range), leaving their fast attack SSN submarines behind to track the opponent's SSBN. By the end of the Cold War, Soviet SSBN submarines no longer ventured too far from their own bases, as they now possessed missiles that could strike US targets from their alongside berths. To keep away from prying eyes from satellites, they submerged in areas within the Anti-Submarine Warfare Defenses of their own harbors, and that was their operational patrol. Russian fast attack SSN submarines like the 'Victor I' and 'Victor II' crossed into the Atlantic Ocean, looking for US SSBN strategic submarines. Also like them, Russian SSGN submarines of the 'Oscar' and 'Papa' classes went looking for American Carrier Task Forces, to tail them.

As the Cold War progressed, these giant underwater fleets became more powerful than any of the land or air-based weapon platforms. Technology combined with complex organizational support to give them strategic capabilities and deployments. The submarine was indeed the weapon of the Cold War just as the aircraft carrier with its embarked squadrons had become the dominant weapon towards the end of the Second World War. Aided by spy satellites and intelligence reports, US and Soviet submarines stalked each other's' forces, monitored each other's' communications, clandestinely took periscope photographs of each other's' harbor and other vital installations, and even landed agents on each other's' shores while constantly targeting each other's cities and vital installations from under the sea with Weapons of Mass Destruction (WMD). Indeed, the peace that prevailed during the Cold War could be attributed to these very submarines and their retaliatory second strike capabilities.

A nuclear war could not be won by either side. Some 20 to 40 underwater collisions (an unofficial claim) took place between opposing submarines during the Cold War that were never reported by either side. Some have been made public much after the Cold War. With the end of the Cold War, and the Soviet Union breaking up, one would imagine that this cat and mouse game would have ended. There was a short lull, but the game continues even to this day between major powers that deploy submarines on strategic, operational, and tactical missions as a part of power projection.

The Falklands/Malvinas War 1982. The Falklands/Malvinas War of 1982 showed up the strength of submarines when properly used, and the weaknesses when used improperly. It also showed up the wisdom of integrating the submarine fully into the overall military force structure, particularly when fewer submarines are deployed. The effect is less if they are operated singly and independently. The importance of constantly feeding them with intelligence, or being used as intelligence platforms themselves to fully realize their potential, was also one of the lessons learnt in the aftermath. This was not followed and hence submarines did not make the impact expected of them during the war. Having said that, the presence of the British submarines exercising *sea control*, and the sinking of the cruiser *Belgrano* by a British submarine resulted in the Argentineans withdrawing their fleet from the battle space and sheltering them in harbor. Submarines of the Argentinean navy were wasted. During this comparatively short war, submarines were once again used to carry out operational – tactical missions.

Indo-Pak War 1971. During the Indo-Pak war of 1971, the Indian Navy deployed submarines off the Makaran Coast and in the northern part of the Bay of Bengal to seek and destroy warships and merchant ships of the Pakistan Navy. They were chartered to carry out operational-tactical missions, without a strategic aim. The orders given to the Indian submarines were to positively identify ships as belonging to Pakistan before attacking them. To make it more complicated, they were also informed that all Pakistan merchantmen were moving about disguised as neutrals.

By tying the hands of the submarines in this manner, this deployment strategy was doomed to failure from the very beginning. Only warships could be attacked easily, but after the missile attacks on Karachi and the Pakistan Navy, the latter did not venture out into deeper waters where submarines could attack them. They chose to anchor their warships in waters too shallow for the Indian Foxtrot class submarines to pursue or attack them. There were no warships of the Pakistan Navy in the Bay of Bengal. Were their Lordships, who were trained initially in the UK (some of them having even participated in World War II), influenced by Article 22 of The London Naval Treaty of 1930 that formally spelled out the guidelines and procedures that a submarine was required to follow when attacking an enemy non-combatant vessel? The Treaty required an attacking submarine to first inform the vessel of its intentions to sink it, and give time and opportunity for the passengers and crew to abandon ship. In the event, all that was achieved was a *blockade* off the Pakistan coast and, if that was the requirement, it was successfully met.

The Turbot Dispute 1995. The Turbot dispute between Canada and Spain in 1995 is a prime example of conventional submarines affecting policy. *HMCS Ojibwa*'s patrol on Georges Bank in 1993 and *HMCS Okanagan*'s patrol on the Grand Banks in 1994 were very successful, while the deterrent value of a submarine presence during the Turbot dispute with Spain and the European Union in 1995 was a decided advantage for Canadian diplomacy.

Post-Cold War Deployment Strategies

China's Maritime Postures. A new power is slowly and noticeably entering the maritime sphere and the world is keenly observing the build-up of PLA (Navy) following that nation's economic growth in leaps and bounds in recent years. The existence of large numbers of submarines in the PLA(N) inventory, and their periodic use in flexing muscles, have definitely had a bearing on postures taken by the US, South Korea, Japan, Taiwan, and ASEAN nations in that region. China undertook massive naval operations that included submarines, on the eve of Presidential elections in Taiwan. USA reacted by deploying a large naval force in the region to negate the effect of Chinese naval operations. At that time, the Chinese PLA (Navy) was of modest size and capabilities. China's deployment of conventional submarines in 1992 swayed foreign ministers of ASEAN attending the 1992 ASEAN Foreign Ministers meeting in Manila to believe that China had ulterior motives, and resulted in their wanting to formulate new policies to deal with China

The Chinese have made their intentions clear. In a departure from earlier policies, they are expanding their field of activity to spread their influence far afield. They have a 'from the land' and 'from the sea' strategy to enter the Indian Ocean arena. Through the land, the 'Silk Route' is being revived. Their warships first entered the Indian Ocean as far back as in 1986 (three ships to Chittagong, Colombo, and Karachi). They are using political, military, and economic efforts to increase their footprints in the Indian Ocean with what has been described as their "string of pearls" policy (getting facilities in ports for dual use – commercial and military). They are already seeking and consolidating bases for their ships and submarines with friendly nations in the Indian Ocean Littoral. In Oct 2014, a 'Song' Class submarine accompanied by a tender ship visited Colombo as a follow up to an earlier visit by one of their submarines. In May 2015, a 'Yuan' Class submarine (with AIP) docked in Karachi for a week. This was the fourth or fifth submarine to dock in ports in the Indian Ocean during 2014-15. The frequency of such forays is only likely to increase. ASEAN countries are watching these disturbing developments and building up their submarine forces to modest levels from scratch. They watch helplessly as China constructs an artificial island in the disputed Spratley-Paracels region in the South China Sea. An India, that hitherto dominated the Indian Ocean albeit with no military intentions, will now find its task cut out to be more alert than ever, and to monitor Chinese activities in the Indian Ocean. India has also augmenting her navy with a better surveillance capability and newer submarines. Tension in these regions is likely to be high in the years to come.

The few examples that have been narrated above highlight the effect some of the various submarine deployments, both during peace and war, and the effects submarines have had as underwater platforms. There is no doubt that SSBNs will continue to be deployed in their deterrence roles as hitherto. SSNs and SSGNs are finding a different role to play. Not all nations are endowed with nuclear submarines in their inventory, and most maritime nations are equipped with conventional submarines of the modern era. What would their deployment strategies be? Smaller navies with only conventional submarines in their inventory are likely to deploy them for limited operational and tactical sorties mid-ocean, and mostly in pursuance of sea denial roles off their own coast or off the coast of their weaker adversaries, - unless they are a part of coalition forces.

'From The Sea' Coalition Deployments. Transitions in the world order towards the end of the last century have resulted in transitions in the way submarines are being deployed. From a bi-polar world we have moved into a uni-polar world of multi-polar collection of interests. Modern trends indicate an increasing tendency for nations to deploy naval forces in a multinational arrangement in regional conflicts. The blue water concept has changed to a littoral warfare, 'from the sea', concept, against a common adversary. Examples are many - Bosnia, Iraq, and Somalia, to name a few. Submarines with land attack cruise missiles are useful participants in such force compositions. Anti-submarine warfare has taken a back seat and littoral warfare has come to the forefront. To take a page out of the US Navy's doctrine, their submarines are getting more and more multi-mission oriented. From strategic intelligence gathering, they have moved to tactical intelligence gathering. Sea Denial, Special Operations, Operations in Support of Surface Ship Groups, and Precision Strikes on land based targets, have now assumed higher priorities. They have converted four of their 'Ohio' Class SSBNs into SSGNs for this very purpose (more on this later).

From all that has been said, we can see that submarine deployment strategies have changed in the last century based on the need of the hour, and advances in technology. They will continue to evolve based on these two factors.

PASSAGE PLANNING

"All submarine deployments should be planned with The Principles of War as the Bible"

- Anon

'Concealment', 'surprise' and the ability to move up and down in the vertical plane in water, are the key advantages submarines have over other platforms of war at sea. While these characteristics make her a potent weapon, their neutralization deprives her of her lethality and renders her vulnerable.

A successful submarine deployment is the culmination of a whole series of preparations ashore before she is sailed out, followed by optimum support at sea, and safe return after the job is done. This is achieved through careful staff work and planning, and formulation of proper orders that are then issued before the submarine's departure. The more meticulous the deployment orders, the more likely that the mission is going to be successful.

The selection of the *aim* of deployment, and the *mission* to be assigned to the submarine, would evolve from what the national or service imperatives are. During peacetime, it could be just for training, work-up, to gather intelligence or information, or to make a presence felt in a certain area. It could be in support of that nation's foreign policy requiring its navy to carry out certain functions in a certain area, or as part of an effort by the Alliance to contain a local conflict between two belligerents in the region. During a conflict where one's own nation and navy are directly involved, the aim and the mission assigned would evolve or derive from the overall aim assigned to the navy, and the part to be played in it by submarines.

Submarines can be deployed on *strategic, operational, tactical, or special missions.* They can be deployed in areas for sea control, or sea denial and on *defensive or offensive missions.* They can be deployed from their bases or from advance bases where they are supported by a submarine tender or its equivalent. Submarines commonly operate singly. On rare occasions they also operate in groups. Sometimes submarines operate in tactical coordinated operations with ships and aircraft, and this has gained in popularity lately. There may be occasions when submarines, though operating independently, are deployed in the vicinity of friendly surface forces and directed to act independently.

It is not uncommon for submarines to be deployed in overcrowded seas – areas where neutral shipping plies regularly. Conventional submarines, depending on the roles they have been designed for, are either deployed on operational or tactical missions, which could be defensive or offensive in nature, involving sea control or sea denial.

A strategic role for a conventional submarine is difficult to visualize, unless she carries strategic weapons (as some indeed have in the past), or operates in restricted waters. However, with progressively increasing miniaturizing of the nuclear warhead on missiles, it is not difficult to imagine a day when states possessing conventional submarines follow Israel's example and give them a strategic role of a limited nature. These deployments would have to be planned carefully before they are executed.

Planning, actually, is the most essential element of *successful submarine deployments*. Both during peacetime and hostilities, submarine deployment routes are planned with a view to: -

Reaching and occupying a selected area of deployment, to carry out an assigned task.

- Mustering for re-deployment to another area.
- Providing support to other submarines or surface units.
- Returning to base after completing the assigned task.
- Proceeding for work up to an assigned area.
- Moving from one base to another.
- Proceeding for refit to another port/base.
- Carrying out any other assignments requiring routing.

Meticulous planning and forethought must precede the deployment of a submarine, with every attention paid to safety factors for ensuring that she gets to her area of deployment safely, and exploiting her positive attributes to the maximum extent possible. She could be assigned direct or indirect routes to and from her area of deployment. Care must be taken to ensure that her route and that of other friendly forces do not crisscross at the same time. If the two are going to operate in very different areas, during different periods, then there is no need to keep each informed of the other's whereabouts.

However, on the off chance that the surface forces are likely to run into the submarine's area during the course of maritime operations, the authorities ashore controlling the submarine may like to draw a *separating line or zone* and inform surface units to keep clear, well in advance. This would imply that the authorities ashore controlling the submarine know, with reasonable accuracy, the whereabouts of the submarine at all times. This is possible only if careful groundwork goes into planning the submarine's deployment, and the deployment orders are carefully worded. One example would be to order the submarine to pass through pre-assigned way points at particular times (give or take up to an hour or two, up or down).

Planning the passage 'to' and 'back' from the allotted area of deployment should take into consideration the following factors: -

- The aim of deployment and the task assigned.
- The weapon mix required to be carried, and the place and time required for embarking the same. The weapon mix would depend on the kind of opposition likely to be encountered in transit, and the type of mission required to be carried out in the deployment area. Since most submarines have options, the optimum and most suitable mix should be arrived at.
- Whether a single submarine or a number of submarines are being deployed. If more than one submarine is being deployed, then their routes should either be spaced apart with a safety margin thrown in, or their movements along the same route staggered, while retaining a measure of freedom of operation for each of them.
- Whether alternate routes are required to be planned for each of the submarines, if more than one is being deployed.
- The requirement to be escorted or 'protected' during certain portions of the transit. (Russian submarines routinely took shelter under their merchant ships and warships when penetrating the GIUK cordon during the Cold War).
- Absence or likely presence of minefields and other obstacles while leaving harbor, and en route.
- The nature of task assigned, and the transit time available. The transit time available would indicate how early the submarine should be sailed, and the speed of advance the submarine should maintain. If there is less time, she would have to take a more direct route and move faster, thereby

possibly compromising on her attributes of 'concealment and surprise' considerably. If more transit time is available, she could make a cautious, silent, and slow approach, or even an indirect approach, to her area of deployment. The speeds of advance may need to be varied for different zones the submarine is expected to pass through. She would have to be very discreet when moving through an enemy submarine probability area and be ready to take on or evade another submarine with very little notice. Planning a low speed of advance through such an area would be prudent. Similar considerations need to be given for arriving at speeds of advance through an area with heavy air surveillance, areas with neutral ships plying, and areas where hostile warships are likely to be present.

- The characteristics of the submarine being deployed, her age and her material state. As stated earlier, each class of submarine would have been built with a specific primary role and a secondary role in mind. When sent on a mission, her positive aspects must be used to advantage, and her shortcomings protected. Her material state would depend on a number of factors like her age, when the last refit was carried out, when she was last refueled (nuclear boats), when her batteries were last replaced (conventional boats), and so on.
- The state of work up of the crew. This varies, and is never constant although it is possible to maintain a 'near high' state of work up on a continuous basis. A new crew takes time to settle down. Even after settling down, there are disturbances by way of transfers 'in' and transfers 'out', some of the personnel away on leave or courses, a percentage of the crew just out of submarine school with brand new submarine badges on their chests and no experience, and so on. The state of work up of the crew will tell you what you can set them to do and what you can expect of them.
- The C4SI (Command, Control, Communications, Computers, Intelligence, & Space) organization available to guide or control the submarine. The organizational set-up, the speed of the system to collect, collate, and analyze

information, the coding system, the system in place to pass timely information to the boat at sea, the time delay between an event happening, receiving the information ashore, and getting it to the submarine at sea, the ability to order re-routing the boat already at sea if considered necessary, etc should be factored into the considerations that go into understanding the effectiveness of the C4SI system in place, and its exploitation.

- The nature of opposition likely to be encountered. Information by way of intelligence, an appreciation of the disposition of opposing forces, their pattern of operations etc, form vital inputs to arrive at the nature of opposition the submarine on transit would be expected to face. The probability of detection by any platform (aircraft, ship, another submarine, or even a neutral, while on transit) would give valuable inputs for her survival considerations (dealt with in another chapter).
- Weather, hydrology, and geographical considerations, and their effect on various options available so that they act most favorably on the departure time and final transit route selected.
- Limitations to the number of Transit routes that can be planned. This would emerge from regional geographical limitations, the number of forces milling around in the area, and other requirements warranting restrictions to be imposed on the number of transit routes that could be planned, and so on.
- Return passage to be selected for turn-around purposes.
- Return passage to be selected in event of damage.
- The Rules Of Engagement (ROE) in force.
- Any other considerations considered essential for passage planning.

Surface forces, moving in a task force or group, have the advantage of flexibility in selection of targets when engaging a 'visible' enemy, a choice of a host of weapons to select from, flexibility in distribution and allocation of firepower, ability to concentrate or disperse, better mobility, and speed. Between the many platforms that are normally widely spaced and form a task force or group, they can monitor much larger areas than submarines. They need space for maneuvering and 'surprise' is hardly an element they could use to advantage, being very visible on the surface of the sea.

Submarine movements are concealed and so when the two are operating in each other's vicinity, it is important to ensure that each does not tread on the others' toes. This is true not only in the area of operations, but also along the transit routes to and from the area of operations, be they direct or indirect. Since submarines do not, as a rule, break radio silence (HF/VHF/UHF) at sea, there should, ordinarily, be no radio communications between the submarine and her surface forces operating in the vicinity as they could be intercepted by the adversary.

Modern submarines communicate with e-mail chat and have separate chat rooms for each communication channel. They are difficult to intercept, and by the time they are, if at all, their relevance would have become history. Some have SHF Satellite Communication facilities for communications. These modern facilities are usable to communicate with others, like in the case of an SSN operating integral with her surface units as part of a carrier task group, or a boat with her submarine launched cruise missiles (SLCMs) operating as a part of coalition operations against targets ashore. There may be rare occasions when they can be asked to transmit and communicate on other than satellite, sonar, or 'Internet' channels. On such rare occasions, the deep sound channel or towed wire antenna could be used, as they offer more discreet options.

The SSBNs would normally all be deployed on strategic missions. Only the United States of America, Britain, France, Russia and China form the exclusive club that possess these submarines, built for strategic (also 'deterrent') roles. There are one or two new entrants knocking at the door. The area of deployment of these submarines is largely based on the range of their ballistic missiles, the targets they are aimed at, and the kind of opposition they are likely to encounter, and has already been covered in an earlier section. During the latter stages of the Cold War the former Soviet Union deployed them close to their homeports and inside very strong anti-submarine cordons, termed by the West as 'bastions', that kept them safe and away from prying, opposing, SSNs. The Americans refer to the SSBNs as 'boomers'. Their assigned tasks, though very important, are mundane and nowhere as exciting as that of the Fast Attack SSNs.

An important step towards planning deployments and routings would be to study the charts that the submarine would be expected to use from the moment she leaves harbor, till she gets back. A wealth of information is available in charts that need to be augmented with further related reading. The longer one stares at charts, the more information one gets. (For example, the undersea terrain in shallow waters where the submarine may be expected to operate may reveal a useful shallow basin that can conceal the submarine and keep her away from prying eyes when she wants to lie low and avoid being detected. Only a careful study of the soundings marked on the chart would reveal such areas).

Simultaneously, all available intelligence inputs that affect submarine deployments must be obtained, analyzed, and studied. There are areas in the oceans where submarine activities are higher than what may be considered to be normal. A submarine deployed in such waters is likely to be confronted with complex terrains, currents, temperatures, weather, and environments wherein ships (and possibly, other submarines) of many nationalities, both friendly and hostile, and even neutral, may be present.

While she can discern between different vessels and identify them with the use of her many sensors that she carries onboard, the limited range of observation that she possesses can sometimes handicap her. (Even the '*Astute*' class SSNs of the Royal Navy that lay claims to being able to detect ships leaving New York harbor from the English Channel – a distance of 3000 miles – will have handicaps by way of identification and lack of clarity in bottom bounce techniques and deep sound channels being used). It is worth considering assisting the submarine through an external source (the submarine controlling authority ashore) feeding her with information that could possibly extend her range of understanding of the environment to many more thousands of square sea miles around her than her sensors would permit, even while she is making her way to her area of deployment.

As a first step, the known presence of all relevant friendly and neutral platforms in the area could be fed to her to make the task of differentiating between 'friend' and 'foe' that much easier. Adequate precautions should be taken to ensure that friendly submarines do not come within detection range of each other while on transit, so as to reduce underwater confusion. (Even if perchance they are likely to do so, a policy or procedure for avoiding mutual interference should be laid down). This, again, is best ensured at the planning stage by an external agency that has a wider picture and can consider all or most of the inputs required that is likely to be useful, sift them, and then route the boats to separate areas for deployment. Command and control of submarines deployed at sea, and communications with them are prerogatives that are best handled by authorities ashore, and this is so in most navies. This has been found to be the preferred option to one of control by fleet commanders or local afloat commanders in the area of operations.

Of course, there are exceptions. One such exception is when a submarine is involved in coordinated or coalition operations with surface and/or air units. In such combined operations, it would be prudent for the authorities ashore to hand over control to the local coordinator for the duration of that operation, and take it back thereafter. Another is in respect of SSNs operating as part of a Carrier Task Force, a procedure in use in the US Navy. As the submarine is habitually a 'listener' so as to maintain her advantage of stealth and surprise for as long as she can, radio communications between the authorities ashore and the submarine would mostly be one-way, with the former doing most of the 'talking', satellite and internet communications being exceptions. These communicative messages would normally be coded and kept to the barest minimum, and would mostly consist of new, required, information not already available with the submarine, and which would assist her in her mission.

Working backwards, before a submarine is put to sea, she should have 'up to date' knowledge of all that she is required to know of the area she would be transiting through on her way to her area of operations, and the area she would be transiting through on her way out of her area of deployment. Armed with this knowledge, she sails out. Additional information, changes in scenario, and any other new inputs received after her departure that the shore authorities feel must be fed to her, should thereafter be conveyed to the platform at sea through the network centric system in use so that a picture larger than what onboard sensors can give her, is made available to the submarine commander from time to time as a part of the information warfare system that is now integral to any form of warfare. A preconceived system would have to be put in place for communicating with submarines so as to ensure that they get messages while they are still useful and not after they have been overtaken by events and the information condemned to obsolescence.

Boats should not be expected to expose themselves to receive messages from other authorities at sea. To ask them to do so would negate the very advantages of secrecy and stealth under which they operate. The system put in place should enable them to receive messages in a dived state. What the submarine is required to do in event of damage, or on completion of her deployment must also be thought of before she is sailed out for her mission. Is she to be replenished and re-deployed? If so, where does she get replenished? - And does she return to the same area or to a different area? For the same mission? - Or on another mission? Do new transit routes have to be planned? Will she require to be escorted or given cover at any stage? All these, and the considerations enumerated for her transit and deployment should again be considered for her post-deployment program in the preparation stage itself, and laid down in her mission orders.

One may argue that this gives very little flexibility in re-deployment of submarines during the course of war, based on the progress of battle. The thumb rule is to give the submarine all types of orders to cater to all possible situations likely to be encountered, as anticipated, and then give only a short, coded, executive order to her at sea for her to switch. This requires very extensive preparation work before the mission orders are made out. This is particularly required when communications are one way, i.e. from shore to sea.

The degree of control exercised by the shore authority over the submarine at sea is a contentious issue and can be debated endlessly with points 'for' and 'against' being spelt out quite strongly. It would largely hinge around the ability of the shore authority to know at any point in time as to where his submarine is / submarines are, with some degree of accuracy, and be able to redeploy them at short notice, when required. This would hinge around the ability and speed of communications between them, among other considerations. The lesser the control exercised by authorities ashore, the more the initiative vested in the submarine commander at sea. Finally, as has been explained at the beginning of this section, it must be remembered throughout the planning stage and when writing the orders for the submarine that the key to successful submarine operations is 'concealment' and 'surprise'. The lesser the number of people involved in the deployment planning process, the better. It is said that the amount of victuals embarked onboard quite often gives away the number of days the submarine is putting out to sea, to inquisitive persons. The weapon outfit being embarked also gives away information that others need not know. All these should be executed away from prying eyes to the extent possible. The sailing time should be concealed. Submarines should normally leave harbor on dark nights while much of the world is sleeping, and dive before they awaken. Their departure and return should be camouflaged in every which way it is possible.

DEPLOYMENT AREAS

"They also serve who plan and wait."

Anon

The area where the submarine is to be deployed is the area where she is required to snoop around or lie in wait for the adversary, and successfully carry out the tasks assigned to her. The area is normally selected by the authorities ashore who have planned her transit route and who have deployed the submarine. However, once the submarine gets to the area of deployment, the initiatives and actions required to be taken to fulfill her mission are hers and hers alone. To borrow a well-known saying, you can take a horse to water, but you cannot make him drink. The drinking has to be done by the horse himself. You can plan the submarine's passage to her area; you can suggest how she is to conduct herself in the area; but how she ultimately behaves and conducts herself in the area should be left to her commanding officer entirely. It is a combination of teamwork: the planning staff should do their homework and select a most favorable area for deployment and get the submarine there safely, while the submarine should use all standard submarine practices and skills that she has trained in, to optimally perform in the assigned area.

This section deals with what the planning staff is advised to do to select the best area of deployment for the submarine. In another section, we will discuss what the submarine, on her part, is required to do. The deployment orders must necessarily be in the broadest of terms, leaving the initiative entirely to the boat, for in the area of deployment the situations and conditions may change and vary at a pace too fast for the authorities ashore to observe, react to, and control. It is for the man on the spot to take the initiative and react to changing and challenging situations.

For the commanding officer of the submarine to succeed, the deployment area should be carefully worked out. An underwater chase for long durations due to an improper deployment is an arduous undertaking for conventional submarines in particular. This is especially true for operations in geographically constrained, or shallow, or restricted waters. Sometimes such a chase can be beyond the submarine's abilities, and the quarry may get away. How does one select the area of deployment? Should the submarine be geographically restricted and ordered to operate between confined latitudes and longitudes? Should the area assigned be in the shape of a circle? Or should it be a square box? Could it be a rectangular box? Should it be of any other shape? How big should the assigned area be? Should it just be a 'barrier line' to patrol along? Will the selected area be too restrictive? Should the submarine be given a free run in a general area without any boundaries confining her movements? Would it be a tall order to expect her to succeed if she is given the run of the mill, seeing that she cannot possibly cover the entire area with her finite sensors and weapons ranges at any given time? Would she hamper the movements of own surface forces in the vicinity, if she is given a free run? What about the presence of neutral forces? What percentage of presence must she ensure in her area? How does one arrive at the optimum solution?

Only SSNs have the ability to chase and run down a victim as they have the speeds and unlimited endurance for it. Conventional submarines can only do short bursts of speeds to chase a victim as they are entirely dependent on the finite power in their batteries. It therefore follows that it is wise to place a conventional submarine in the *expected path* of the adversary so that she can lie in wait and go for her objective when the adversary comes within striking range. Geographical restrictions that can be advantageously used must be taken into consideration. But all this is jumping the gun. Let us go back to the beginning and look at the considerations for selection of the area where the boat is to be deployed.

All of the factors that were taken into consideration when planning the submarine's passage to and from the area of deployment must, once again be considered by the authorities ashore when **planning and selecting the Area for Deployment** for the submarine. In addition, there are some other factors to be considered. The submarine's capabilities, the nature of the tasks and the degree of coverage required in a particular area will determine the dimensions of the area and whether only one submarine is required or whether more than one submarine needs to be deployed. In geographically restricted waters one submarine may suffice, but in open ocean and seas, more than one submarine may have to be deployed. This is the missile age, and submarines carry missiles that can be launched from their torpedo tubes, or from separate, dedicated, launchers.

The aim of a deployment, whether for clandestine operations, intelligence gathering, sea denial, or sea control, must ensure that when required, the weapons onboard, when released, find their targets. Selection of the area for deployment must, thus, **take sensors and weapon ranges with respect to targets into consideration**.

The next factor to be considered is the **'percentage presence'** required of the submarine in the deployment area. As the terminology suggests, 'percentage presence' would be the number of hours the submarine would be required to be in the deployment area continuously, over a period of a hundred hours. If the answer is 'all of hundred hours' then the extent of deployment area selected would have to be restricted to just a little over the area she can keep constant surveillance on, and get to the extremities of the area in time to get her quarry. To cover a larger area, the preferred option would best be to deploy a nuclear attack submarine with the necessary speed advantage, if such a submarine is available in the inventory. If the dimensions of the deployment area are to be increased, then a lower 'percentage presence' may have to be accepted when only one conventional boat is deployed. Looked at in another way, the 'percentage presence' would be an important factor to be considered when arriving at the dimensions of the 'Area for Deployment'. If the area is too large for one conventional submarine, and a nuclear attack submarine is not available, the alternative would be to deploy more than one conventional submarine in the area, with a safety separating corridor between them, to avoid mutual interference.

If a conventional submarine is to patrol in a deployment area for a hundred out of hundred hours, she would have to charge her batteries in the area itself. To do this, she would have to come up to periscope depth and run her diesels – a very vulnerable and noisy situation that may lead to her presence being compromised with or without her being aware of it. This brings in the issue of the **'Indiscretion Rate'** of the submarine. The 'Indiscretion Rate' is the ratio of the time needed to remain at periscope depth to recharge the batteries and air, and the total operating time. In the area of deployment, the Indiscretion Rate should be as low as possible so as to maintain concealment and surprise, and ensure survivability. To lower the Indiscretion Rate to as low a figure as possible in the deployment area, a conventional boat is advised to move away to a safe haven some distance from the area of deployment, to periodically re-charge her batteries, and then come back. She would, thus, only absent herself for the duration of battery charge and period of transit to/from the battery charging area. This would result in a lower percentage presence, but a better indiscretion rate in the area of deployment.

How 'low' would also depend on the distance away the haven for battery charging is from the deployment area. If the percentage presence is to be maintained as near to a 'hundred' as possible while reducing the Indiscretion Rate to the lowest possible figure, two or more submarines could be deployed for the same task in the same area with one of them in the area always (exchanging places), and the other/s remaining close at hand, charging batteries or transiting to/from the area of deployment and the haven for battery charging. This would have to be a well-coordinated plan as it is complex, and could involve the need for submarines to discreetly communicate with each other. It is workable.

It should be standard submarine practice to work up a submarine crew gradually through various stages before the boat can be considered to be a 'front line' boat, capable of being deployed anywhere and for any role. Sometimes, a boat not yet ready for front line service may have to be deployed because of the immediacy of the requirement. The **characteristics of the submarine to be deployed, and her material state** must next draw attention and be matched against the type of opposition she is expected to encounter. Equally important is the **state of work up of the crew**.

The C4SI facilities available with the submarine operating authorities ashore, the submarine, and possible friendly surface forces operating in the vicinity, must then draw attention when selecting the deployment area. It must be the endeavor of the authorities ashore to constantly feed the deployed submarine with up to date information relevant to her functioning in the area. This gives her a wider picture and leads to better decision-making on her part. Getting that information to her too late for her use is of no consequence. For example, let us assume that the deployment area has been selected after receiving intelligence inputs about the intended path of the adversary. Subsequent to the submarine's departure, it is learnt that the adversary is going to take another route. The timely redeployment of the boat can place her in the new path of the adversary. If the information is conveyed too late, the boat will miss her adversary.

This is a difficult task as the boat is going to be mostly under water and underwater communications are dependent on a host of local factors in the vicinity of the submarine. Take another scenario: our own deployed surface forces need to suddenly move through the submarine's deployment area – a situation not foreseen when forces were initially deployed. It happens. The information must get to the submarine before the forces move across, and not after. If C4SI facilities available do not afford these flexibilities, the deployment area selected must necessarily be affected.

The **weather, hydrological, and geographical considerations** in the area of deployment, across the entire period of intended deployment must be studied before arriving at the shape and extent of the area of deployment. There are areas with underwater currents, 'up-wellings', extensive bio-luminescence (as a result of presence of marine biological creatures), large rivers flowing into the sea whose effects can be felt for miles out into the ocean, etc. etc. These could have an effect on the submarine deployed in such areas. Bad weather and rough seas may make it difficult for a conventional boat to come up to periscope depth and charge batteries. Hydrology may either give the submarine or anti-submarine forces an advantage in the area of deployment. Shallow patches, underwater pinnacles, coral reefs, floating icebergs, and islets in the region may hinder deployments. These are all considerations to be taken when arriving at the optimum area for submarine deployment.

Mention was made earlier about a **haven** for the conventional submarine to **charge her batteries** in. Mention was also made of whether one or more than one submarine needs to be deployed in an area of interest. The selection of the former is dependent on the decision taken in respect of the latter. In the latter case, more than one haven may have to be assigned. Alternatively, if the same haven is to be used by more than one submarine, it must be ensured that both are not present in the same haven at the same time. The nature of mission and the acceptable period of absence from the deployment area will have a bearing on the selection of the haven. If it is too far away, the period of absence from the deployment area will be that much longer. On the other hand, depending on the adversary's anti-submarine warfare capability, the haven may just have to be located at some distance away. Geographical considerations, and the nature of opposition, will play an equally important part. Selection will also depend on the characteristics of the boat, and the age of batteries. An older set of batteries onboard will warrant more frequent charges as they will discharge faster. The period of absence from the area will be that much more. Weather, hydrology, and geography will have a bearing on the selection and location of a haven for charging batteries too. Selection of the haven for battery charging applies only to conventional boats and does not obviously apply to nuclear propelled submarines.

THE ARCTIC & ATLANTIC OCEANS

"The North Atlantic is a cruel and unforgiving body of water. There's no land up there in the Arctic. It's all ice. Submarines go under there all the time."

Darlene Bailey

Seventy percent of the world's surface is covered by water. From time immemorial mariners have been plying this vast expanse of water. From coastal ventures to trans-oceanic adventures, with the dawn of the sail, the sea provided a means of trade as also a means of carrying men and material across from one shore to another to conquer other lands for their country and king. Fighting navies were created and maintained by maritime nations, either to defend themselves or to fulfill their more adventurous designs. They remain to this day.

The submarine, as an *underwater fighting platform*, first made its successful appearance in 1864 when Horace Lawson Hunley built one for the Confederate army during the American Civil War. Armed with a 90-pound charge on a long pole, the submarine – named after its inventor – attacked and sank a Federal steam sloop, USS Housatonic, at the entrance to Charleston Harbor. With that sinking, the *Hunley* proved that a submarine could be valuable as a weapon of war at sea. Since then, submarines in various forms, shapes, and sizes, have been a part of the inventory of powerful and developing navies. They serve *strategic, operational,* as well as *tactical* interests of nations and navies, as formidable and effective platforms of war and are used for *sea control* as well as for *sea denial.* They (SSNs) also escort Carrier Task Forces as a matter of routine, and form a part of coalition forces for 'from the sea' operations. In the past, they have achieved seemingly impossible tasks with great success. The heroics of Günter Prien in Scapa Flow, or the X-boats attack on the Tirpitz in the Scandinavian fjords during the World War II are enduring testimonies of the submarine's prowess.

The geographical areas of interest of global powers for deploying submarines envelope almost the entire water surface, with particular interests in certain specific regions where '**SSBN**s' (Strategic Submarines **B**allistic **N**uclear, or Ship Submersible **B**allistic **N**uclear), that form the third and elusive arm of the nuclear 'triad', are deployed. Also of interest are the areas where Fast Attack

Submarines (SSNs), with or without surface forces and airborne support, are deployed. These deployments usually serve the purpose of maintaining a permanent vigil or periodic presence in the interest of a military balance or in support of self-interests. This was the scenario during the Cold War that ended in the late 1980s. It has not changed with the end of that Cold War.

Apart from global powers policing the oceans with their submarines and surface fleets, there are nations who acquire and deploy them in support of military alliances they are a part of. There are others without adversaries that maintain and operate them for their deterrence value. Nations, that have long-standing differences and bones of contention, leading to periodic skirmishes, maintain force levels that often lead to an 'arms race' they can ill afford. The maritime component includes submarines.

In all the waters around the globe, there are areas of interest where submarines are deployed. All such areas form *arenas* of interest for undersea warfare. Some of them are of a permanent nature, some of periodic interest, and some of interest only during actual hostilities. For the sake of classification, these arenas may be broadly split into three different classes, and referred to as *'very high' submarine probable areas, 'high' submarine probable areas, and submarine probable areas.*

A '*very high*' submarine probable area would be one where almost a **constant presence** of a submarine, or submarines, may be expected. For example, the areas where strategic submarines are deployed and operate, and where they are relieved on task by their respective successors, would be areas of constant presence and therefore very high submarine probable areas. During hostilities, the approaches to both own and enemy harbors would be '*very high' submarine probable areas* even if only one of the two opposing forces has submarines. They can be deployed off the enemy harbors on offensive roles and off own harbors on defensive roles.

Areas where **periodic presence** of submarines may be expected, may be defined as *'high submarine' probable areas.* Such areas may include approaches to narrow straits or passages through which ships have to, and do, pass on their way to their destinations with periodic regularity. Submarines find it attractive to lie in wait for their quarry near these very straits or narrows. The Greenland – Iceland – UK (GIUK) gap between the Arctic and the North Atlantic Ocean, the

Strait of Gibraltar between the Mediterranean Sea and the Atlantic, the Bosporus and Dardanelles Straits that link the Black Sea to the Mediterranean Sea, the Skagerrak and the Kattegat between the Baltic and North Seas, the Malacca Straits and the Sunda Straits in the Indian Ocean, the Bering Straits between the Arctic and the Pacific Ocean, and similar such areas would be some of the examples of *'high' submarine probable areas.*

During peacetime, each country possessing submarines, designates areas off their coast as submarine exercise areas and promulgates the same so that others passing in the vicinity can steer clear of the area. Submarine exercises are conducted occasionally in these areas and these can therefore be referred to as areas of **occasional presence** and therefore *submarine probable areas.* During hostilities, undeclared or declared wars, the three classifications would be arrived at after an assessment of the military situation, and locations could be altogether different. A look at each of the oceans would show us some of the likely areas that could be termed as possible arenas for undersea operations.

The **Antarctica** being a solid mass of land with ice covering most of the region for the major part of the year has been of no particular interest to any nation for submarine operations to date other than for marine research in the seas around it, conducted by surface ships. One can, therefore, not define any of its adjunct seas as submarine probability areas. If, however, submarines are required to support any clandestine activities on the mainland or in the islands that border it, a classification could be assigned to such areas.

The **Arctic Ocean** is the world's smallest ocean and covers an area of about 5,500,000 square miles. Roughly circular in shape and extending in depth down to a maximum of 1500 to 2000 fathoms (one fathom = six feet) in some areas, it is bisected into two portions by the subterranean Lomonosov Ridge. Much of the area is shallower than 1000 fathoms in depth. The main feature of this ocean with its surrounding seas is that it is almost entirely covered with sea-ice in winters, with very little let up in summers. Floating icebergs abound in the Arctic Circle in various sizes and shapes, particularly during the summer months. The Arctic Ocean saw some significant submarine action during World War II. Allied shipping formed Arctic convoys from Iceland to the Kola Peninsula and from Scotland across the Norwegian Sea to Murmansk and on to Archangel, across the White

Sea. Empty ships were convoyed back. They moved along the edges of the winter ice limit. The Germans deployed submarines and aircraft against them with notable success. The fate of convoys PQ 17, PQ 18 and QP 14 are well-documented examples.

Conventional submarines do venture inside the fringes of ice limits of this ocean on operational missions during peacetime. With the thick layer of sea-ice on top, they can continue to operate under water only so long as their battery power lasts. They have to move out and away from under the sea-ice, or locate *polynyas* in time, to come up to recharge batteries. Modern conventional submarines possessing Air Independent Propulsion Systems (AIPS) are able to extend their stay under the ice for a while longer, but still have to find the time and place to recharge batteries. *Polynyas* are thinner (not more than three and a half feet) layers of sea-ice amidst the thick layers, and their presence is more predominant during summer months. It is possible to locate them by looking up for bright light spots from under the sea-ice with suitable optical and other instruments onboard, as these sparsely spread thin layers obviously admit more light underwater than the vast thick layers of sea-ice all around. In summer months, this area of the midnight sun makes polynyas easier to find. In winter months there is no light and these thin layers have to be located by other sensors and instrument onboard the submarine. They are fewer and harder to find in winter as the ice is thicker. With hardened fins or sails, submarines are able to push upwards and break through *polynyas*, and surface. Operating with the sea-ice above, the boat is concealed from prying eyes and able to move about with less chances of detection by satellites, surface, and air units. Occasionally, sound channels that largely exist at shallow depths just below the ice, give extended ranges of acoustic detection that assist them in their hunting role, but the hunted (submarines) also have the same advantage. Looked at in another way, the reaction times available to both are extended.

Since the advent of nuclear powered submarines, the necessity for these boats to surface to charge batteries no longer exists. Countries having the luxury of possessing both types of submarines preferentially opt for the nuclear submarine for under the ice operations. The US Navy and the Royal Navy have denied themselves an option by doing away with conventional boats altogether. To show her endurance and capability to the world, and to the Soviet Union in particular, the world's first nuclear propelled submarine, *USS Nautilus*, transited the Arctic Ocean and through the North Pole under the ice in August 1958, albeit after two unsuccessful earlier attempts. Not to be left behind, the Soviets soon pioneered their Arctic expeditions under ice with their *'November'* Class nuclear powered attack submarines (SSNs), fourteen of which were built commencing from 1959. Many others have followed since. Patrols in the Arctic are largely undertaken by nuclear powered SSNs and diesel electric SSKs (conventional anti-submarine submarines) who hunt each other and any SSBNs lurking in the vicinity.

The approaches to the Arctic Ocean are mainly through the Bering Straits from the Pacific Ocean, from the seas between Greenland, Iceland, and the United Kingdom (GIUK) from the Atlantic Ocean, and from the North and Norwegian Seas. Submarines from the Arctic ports of Russia have to transit through these 'gaps' to make their way to and from the warmer waters of the Pacific and Atlantic Oceans. Others who want to deploy submarines in the Arctic have also to transit through these relatively 'restricted' or 'choke points'. These approaches are *'high' submarine probability areas* because of the periodic, but ensured, movement of submarines through these areas.

The fringe areas of the Arctic winter, and summer ice limits, may also be termed as 'high' submarine probability areas, particularly off the Russian northern coast where a longer presence may be expected, as these are convenient areas for mission deployment. The approaches to all Russian Northern Fleet naval bases in the Arctic are *'very high' submarine probability areas.* The North Atlantic Ocean has been used extensively ever since the days of sail for trade and transportation of personnel between Europe and America and remained the only link between the 'new' and 'old' worlds with regular shipping routes, till trans-Atlantic flights augmented them. It was also the center of many campaigns of undersea warfare during World Wars I and II. In fact, it is mainly out of the experiences gained in this arena that anti-submarine and submarine warfare has developed to its present state.

The seas between Greenland and Iceland (Denmark Strait), and Iceland and the British Isles, will remain a *'high' submarine probability area* in peace, and during any future hostilities in this part of the world. All 'approaches' to naval bases beyond the 50 meters sounding line may be deemed as *'high' submarine probability areas* so long as that country to which these bases belong, has a potential adversary who possesses submarines. On the Western Atlantic seaboard of the United States of America, the approaches to the bases in Norfolk (Virginia), Kings Bay (Georgia), Groton (Connecticut – near New York City – New London seaport), and Charleston (South Carolina), from where the US Navy's submarines operate, may be deemed to be *'very high' submarine probability areas*, where Russian and Chinese SSN submarines may be expected to lurk outside the range of static defensive seabed sensors.

The Eastern seaboard of the Atlantic has strategic submarines of the SSBN, and SSN variety based in Scotland and France. The approaches to their bases may be deemed to be *'very high' submarine probability areas*. The approaches to other naval bases of UK, France, Spain and Portugal on the Eastern Atlantic seaboard may be deemed to be submarine probability areas in the present environment of peace and tranquility that prevails in that part of the world.

The **South Atlantic Ocean** has been an area of relative quiet for two reasons: the Suez Canal and the Panama Canal have ensured that shipping in this area is scant, and traditional adversaries find it too far away from their respective bases to conduct warfare in that area that will have any meaningful results. Except for the Falklands/Malvinas conflict between Argentina and Britain in 1982 when a British nuclear submarine, HMS Conqueror, sank the Argentinean light cruiser, *General Belgrano*, with conventional torpedoes, there have been no conflicts involving submarines in this area post World War II. South American countries are generally at peace with one another, though not always necessarily without tensions.

To therefore define *'very high'* and *'high' submarine probability areas* in the South Atlantic would be difficult. US and British submarines do make periodic forays to the West Indies, the South Atlantic, and Falkland Islands quite freely. However, on the western seaboard of this ocean, traditionally rival navies of Argentina and Brazil both have submarines and the approaches to their submarine bases would be *'high' submarine probability areas*. They would also have *submarine exercise areas* of their own where periodic presence of submarines may be expected. South Africa, on the other hand, has submarines of its own without any potential adversary with a maritime capability – more as a deterrent or 'threat in being' than anything else. They have their submarine exercise areas that are promulgated and shown on charts relevant to that area. The North Atlantic may continue to be seen as a hotbed for peace-time submarine operations in the foreseeable future. There are hotbeds also in the seas that are linked to the Atlantic Ocean.

The **Mediterranean Sea** covers an area of about 965,000 square miles with the 14 kilometers wide Strait of Gibraltar providing an opening into the Atlantic Ocean, and the Suez Canal providing the only other opening - into the Indian Ocean. During different stages of civilization, it has been called by different names – Eurafrican Mediterranean Sea, European Mediterranean Sea, Mare Nostrum ('our sea' in Latin), the Sea of the Atlantic surrounded by the Mediterranean Region, so on and so forth.

The Romans coined the word **'Mediterranean'** meaning 'Middle of the Earth'. It has been the most traversed sea of all since man began to journey over water. The Mediterranean Sea is a combination of many smaller seas – the Tyrrhenian Sea, the Adriatic Sea, the Ionian Sea, and the Aegean Sea. Flowing into it from the east is the Black Sea. The Mediterranean Sea was a hotbed for submarine operations during World War II. Even today, Spain, France, Italy, Albania, Greece, Turkey, Syria, Israel, Egypt, and Libya, all operate submarines in the Mediterranean Sea, making it a very busy arena for even peacetime submarine operations. There exist many *submarine probable areas* as a consequence thereof.

The **Black Sea**, with an area of a little more than 165,000 square miles and a maximum depth of 2,200 meters, has a positive water outflow and pours out around 300 Km^3 of water per year into the Mediterranean Sea through the Bosporus Strait. It was the abode of the Soviet Black Sea Fleet (mainly) till the breakup of the Soviet Union, and now has the Bulgarian, Russian, Ukrainian, and Turkish submarines operating in the region. There are many *submarine probable areas* adjoining each of these respective countries. There is stability and peace of a disquiet nature, and submarine operations are limited to training and peacetime missions. It is an enclosed sea and, therefore, very much at the mercy of those controlling the Dardanelles and the Bosporus Strait, to make an entry into the Mediterranean Sea.

PACIFIC & INDIAN OCEANS

"In one drop of water are found all the secrets of all the oceans."
Kahlil Gibran

The Pacific Ocean.

The largest of the Oceans, the Pacific Ocean, covers a third of the entire world's surface. It is also the deepest of the oceans with an average depth of around 4300 meters. The ocean is dotted with innumerable islands, small and big, most of which have been spared the negative aspects of modern civilization. Some of the more strategically placed ones have been occupied by navies that have chosen to dominate the ocean by using these islands as their forward pivotal bases.

The Pacific Ocean, like the North Atlantic Ocean, saw major activities by US and Japanese submarines during World War II that got triggered by the infamous Japanese attack on Pearl Harbor on 07 December 1941. It drew the United States into the war. The different strategies in submarine deployment adopted by the two sides, and consequent successes gained, are discussed in another chapter. Post World War II, the Cold War saw an entirely different pattern of submarine deployment, with deterrence, anti-submarine warfare, and special operations forming the main missions. Of major current interest from submarine operations' aspect, starting from the north, we have the Bering Straits that form the sole gateway for Russian submarines emerging from the cold Arctic Ocean for forays into the warmer Pacific Ocean. It is anybody's guess as to how many submarines, from both sides (mostly SSNs chasing and tailing each other), had been milling around these waters during the Cold War. They still do, but with less intensity. The approaches to this Strait will always remain a *'high' submarine probability area.*

The **Sea of Okhotsk** and **the Sea of Japan** were the home waters of the erstwhile Soviet Pacific Fleet, and are now the training waters for Russian submarines based in Vladivostok and Kamchatka. The Island of Sakhalin is a long-standing bone of contention between Japan and Russia. Clandestine missions and patrols by submarines of both countries around this island continue. At the time of writing this book, Japan has some 25 modern, sophisticated, indigenously-built conventional submarines of *the 'Harushio', 'Oyashio', and 'Soryu'*

Classes. Some of them deploy in these areas. One can safely conclude that the sea around Sakhalin Island is a 'high' submarine probability area.

Also, milling about in the neighborhood, in the Sea of Japan and the Yellow Sea, are submarines from opposing North and South Korea. Their submarines frequently flex their muscles in each other's waters and, over the post-1950s, have had more than one incident which has led to diplomatic wrangles that have drawn the attention of the world. The sinking of a South Korean warship in mid-2010 by a torpedo fired by a North Korean submarine in peace-time conditions, as alleged by South Korea, raised the diplomatic wrangle to a new high. North Korea has succeeded in launching a ballistic missile from their conventional submarine which has raised quite a few eyebrows in the region. In August 2015, in a show of strength, over 50 of North Korea's 60 plus submarines – both new and old – put out to sea simultaneously with both the US Navy and the South Korean Navy disturbingly unable to locate the whereabouts of a good many of them for quite a while. Tensions were again at a high as a consequence thereof. With constant unrest between the two Koreas despite ongoing and continuous efforts to bring about a rapprochement interspersed with short periods of belligerence, the waters around the Koreas, and the Yellow Sea (it also has the Chinese submarine bases at Qingdao, Huludao, Lushun, etc) can be deemed to be *'high' submarine probability areas.*

The Yellow Sea, the **East China Sea**, and the northern part of the **South China Sea** had been dominated by a very large number of 'R' Class (Russian design) submarines of the Chinese Navy [PLA (N)] during the latter half of the last century. They are all now very dated. The whole submarine fleet is going through a process of modernization with SSBNs, SSNs and modern conventional submarines (the '*Son Won – II/Yuan' Class, the 'Cheng Bogo' Class*, both based on German designs) progressively replacing their dated submarines. The conventional submarine force (a little over 60 boats) is now capable of being split on an as-required basis for flexible deployments with their 'North Sea Fleet', 'The East Sea Fleet', and their 'South Sea Fleet'.

The base in Jiangghezhuang, near Qingdao, has a waterway entrance to underground facility for 'Xia' type SSBNs. According to Chinese stated policy, they are all being deployed for 'defensive deterrence'.

Major submarine bases for their conventional fleet are in Qingdao (HQ), Huludao (northern-most part of the Yellow Sea), , Guzhen bay, Lushun (near Dalian), and Xiaopingdao. Their surface fleet earlier formed the 'East Sea Fleet', with headquarters in Ningbo. Their 'Amphibious forces' still form the bulk of their 'South Sea Fleet', and confront Chinese Taipei (Taiwan).

The submarine probability areas are not too difficult to discern in their immediate seas. Nevertheless, with the US Navy's almost constant presence in the region in support of Taiwan, the areas around Taiwan are definitely *'high' submarine probable areas.*

The **South China Sea** is going to be a *'very' high submarine probability area* for the foreseeable future. In the years to come, to expect a concentration of warships and submarines of different nations in the South China Sea, and particularly around the Paracel Islands claimed by China, Taiwan, and Viet Nam, and the Spratly Islands claimed by Brunei, China, Malaysia, the Philippines, Taiwan, and Viet Nam, would not be out of place. They are oil-bearing regions and hence the special interest. China is making man-made islands in the region by dredging sand from a depth of around 300 feet or so, to build fortified military bases, particularly around Subi and Mischief Reefs. At the time of writing this book, she has already reclaimed some 2000 acres of land, turning mere sandbars into islands equipped with airfields, ports, and lighthouses. The move has raised hackles in the region. Apart from PLA (N) submarines, newly acquired submarines of ASEAN nations are going to exert their presence there. Between them the ASEAN nations have, at the time of writing this book, over twenty modern conventional submarines available to counter a perceived threat from China and its declared intent. That is a formidable force level that cannot be ignored, and is being further elaborated upon, below. However, co-ordination of a very high level will have to be executed to avoid mutual interference in the pursuit of common interests.

Conventional submarines that belong to Taiwan operate in the Straits of Formosa and the South China Sea. The Taiwanese Navy operate two Dutch *'Zwaardvis'* Class conventional submarines. They have replaced their earlier Dutch *'Dolfijn'* Class submarines. They are on the lookout to augment their submarine fleet further.

The Singapore Navy has acquired new AIP-fitted submarines of the *'Archer'* Class from Sweden in 2005 to augment their *'Challenger'* Class submarines acquired in the 1990s from the same country.

The Indonesian Navy is in the process of acquiring three improved *'Kilo'* Class submarines and three improved *'Type 209'* Class German submarines to augment or replace their *'Cakra'* Class submarines acquired in the 1980s from Germany.

In Sep 2009, the first of two French *'Scorpene'* Class submarines joined the Malaysian Navy. To also operate them in the Indian Ocean, they have created a new submarine base in Pulao (Langkawi), on their west coast. They already have a base to operate submarines in the South China Sea.

The Vietnamese navy will have six improved *'Kilo'* Class submarines with *'Klub'* cruise missiles in its inventory by 2016.

Thus it can be seen that the entire western Pacific offshore belt is peppered with areas of interest for submarine operations.

Further down south, the Australian Navy has a modest squadron of indigenously built *'Collins'* Class submarines that have troubled them endlessly with their performance. The Australians have shown interest in acquiring the *'Soryu'* Class submarines from Japan as replacements.

The eastern limits of the ocean are dominated by submarines of the United States navy that has submarine bases in Pearl Harbor (Hawaii), Ballast Point (San Diego – California), and Bangor in Washington (Puget's Sound in Kit sap peninsula), for SSBNs, SSNs, and the Deep Submergence Rescue Vessel (DSRV). A forward deployment base also exists in the island of Guam.

The Indian Ocean.

With an area of around 74,000,000 square kilometers, the Indian Ocean covers a seventh of the world's surface area. It has an average depth of around 3900 meters. Bounded mostly by littoral states that were colonized for years and still recovering from the after effects, prosperity is still a far cry from reality. It has been named after India for very good reasons. By inverting the chart of the Ocean and looking at it, it can be seen that the projecting landmass of the Indian sub-continent in the center dominates the sea, thus stressing the fact that the Indian Ocean is important to India, and vice versa. Littoral States of the Indian Ocean are aplenty. In comparison, few possess navies, and of these, most are 'brown water' or coastal navies. Yet, in one sense, the Indian Ocean is the most important of all the Oceans, because of the large fossil fuel reserves in the Arabian States around the Persian Gulf, in the waters of the North Arabian Sea, and in Indonesia. Many developed nations have a naval presence in the North Arabian Sea to safeguard their oil interests. In the process, they have acquired forward support bases in the Indian Ocean.

The United States of America has signed a lease agreement with the UK government and taken over the island of Diego Garcia in the Chagos Archipelago that strategically positions them right in the middle of the ocean from where they support their ships, submarines, and aircraft operating across the Indian Ocean. The British continue to have a garrison there. The French Navy has former colonies off Africa, and with continued good relations, is able to support their ships and submarines in Djibouti, Mayotte Island in the Comoros group of islands, Reunion, Mauritius, Seychelles, Malagasy, and other ports. The Russian support comes from Viet Nam where they have suitable facilities in Camh Ranh Bay, and from Yemen's island of Socotra at the mouth of the Gulf of Aden (temporarily suspended because of the unrest there). Others have similar facilities in different parts of the Indian Ocean.

A new entrant is quietly developing facilities in African countries abutting the Indian Ocean, in Pakistan, in Bangladesh, in Myanmar, and wooing other littoral states in support of its expanding interests in the Indian Ocean, and that is China. It is only recently that the PLA (N) has begun to move away from a 'brown water' syndrome to a 'blue water' image with the acquisition of an aircraft carrier, indigenous modern warships, and submarines. At the time of writing this book, their conventional AIP fitted submarines have been seen in 'blue waters', in the Indian Ocean. It is believed that their SSNs are also making forays far and wide. The PLA (Navy)'s 'blue water' ambitions are well underway, and the Chinese are expected to move and operate their submarines further and further afield. This is borne out by their quest for support bases in the Indian Ocean Littoral (IOR). They already have a toe-hold in the Pakistan port of Gwadar and a warship presence off distant Somalia, to counter piracy – a sure indication of their intentions in the years to come.

The strongest *regional* navy continues to be the Indian Navy for the moment, but she has no pretensions of operating as a strong regional power. India's coastline being vast, the navy looks after her maritime interests. She has a dormant "look east" policy that takes her into the South China Sea working closely with the US Navy and the Japanese Maritime Defense Force. She is also working closer with the ASEAN states. She has acquired berthing rights for her ships in Viet Nam, and is helping that nation with offshore oil and gas explorations. (The Chinese term this and other Indian activities in the Indian Ocean as the Indian 'Iron Curtain' strategy). India has a formidable battery of conventional submarines and is on the way to making her own fleet of nuclear power propelled boats to complete the third arm of her 'triad'. At the time of writing this book, she has, on lease, a Russian *'Akula II'* SSN submarine that she is operating, and her first nuclear propelled SSGN is in the process of being inducted into the navy. The Indian navy operates nine *'Kilo'* Class, and four *'Shishumar'* Class conventional submarines, of German design, in the Indian Ocean. In the pipeline, is the French *'Scorpene'* submarine, being built locally.

It is no secret that the world's coveted and dwindling reserves of oil are located in the Middle East in large quantities, and countries are vying with each other to buy and take away as much of this oil as possible from very willing owners who find their very prosperity in the sale of this commodity. This results in heavy traffic of tankers to and from the **Persian Gulf** into, and out of, the Indian Ocean. Westwards, the oil is transported through the Red Sea and the Suez Canal by tankers that do not exceed the draught restrictions imposed in those waters. The heavier ones go round the Cape of Good Hope. Eastwards, the movement is mainly through the Malacca Straits with the deeper draught vessels moving through the deep Sunda Straits of Indonesia.

As the reserves of oil dwindle – and indications are that they must head that way – the price of oil will go beyond the purse of many, leaving only the very rich nations affording the soaring costs to take this black gold away. In a uni-polar world, the lion's share will be taken by the dominating nation by whatever means it decides to choose. This has already been demonstrated. This black gold is so prized that nations, purchasing and transporting it away from the Middle East, find it comforting to afford their tankers some protection by placing their warships in the region. While the excuse given is *"in the interest of peace and stability in the region"*, or *"the threat of piracy in the region"*, plainly speaking they are there to pressurize local littorals to behave the way they are expected to, and to safely take the oil away.

Thus one finds a host of warships, of different nationalities, of nations who form a part of the Indian Ocean littoral States, of nations far removed, in consonance with their allies or independently, all contributing to this very big melee that exists in the Indian Ocean region – in the **North Arabian Sea** in particular. Either through their allies, or independently through the bases already held by them in the region, the bigger navies that belong to nations not located in the Indian Ocean operate with considerable logistic support and back-up from the littoral states of the Indian Ocean Region (IOR).

Among the littoral states in the region, only Iran operates three *'Kilo'* Class submarines, mostly in the **Persian Gulf**, and Pakistan deploys her five French *'Agosta'* Class submarines off her Makaran coast and the Indian coast. Pakistan is in the process of acquiring eight *'Yuan'* Class submarines from China, and is building her first nuclear propelled submarine. For the foreseeable future, till the oil reserves in the Persian Gulf region lasts, the north Arabian Sea will be an area of *high submarine probability area.*

Lastly, historical differences between nations in this region, followed by belligerent postures off and on, result in their respective navies constantly acquiring or manufacturing state of the art naval platforms and weaponry in an arms race that shows no signs of abating, but is thankfully restricted by limited purchasing powers.

It is in this environment that submarines operate, when deployed in the Indian Ocean region. The movement of multifarious platforms of a number of nations, and geo-political oddities of the region, combine to make undersea operations in the Indian Ocean quite different to elsewhere in the world in present times, and makes it a challenging and complex task.

ACTIONS BY THE SUBMARINE

"No thought of flight, none of retreat, no unbecoming deed that argued fear; each on himself relied, as only in his arm the moment of victory."
- Milton: Paradise Lost

Having generally identified potential deployment areas for submarines across the globe, let us see what submarines can do when deployed in such areas. Despite being a potent and formidable platform, much of the submarine's effectiveness is dictated by the type of sensors and ordnance she carries, her age, her material state, and the experience and state of work up of the crew. A submarine just out of refit and modernization is materially more reliable than one that has finished her operating cycle and is in line, waiting for a refit and modernization. A brand new submarine is more contemporary and menacing than a twenty year old submarine. The difference between a submarine that has just come out of refit but not yet worked up as a crew, and a fully worked up operational submarine, is like the difference between chalk and cheese. A well worked up and experienced crew can handle an old boat better than a raw crew can handle a new boat. Unfortunately, while peacetime deployments are timed with the state of work up and material state of boats, operational deployments suddenly crop up and need not coincide with the crew and submarine being in an optimum state. A requirement to send a boat out on an operational sortie can come up at any time. When sending a boat out on an important and rather sudden sortie, it is advisable to ensure that the key personnel manning her are well trained, well worked up, and experienced, and the substitutes being embarked have served on the boat earlier and are familiar with the boat. This is advisable as her every performance will depend on the actions taken by the crew onboard. In preceding chapters, having looked at all the considerations that must go into the operating authority ashore planning sorties and taking the lethal platform to her deployment area, let us now look at all that the submarine must do while on passage, and on arrival in her deployment area.

Leaving harbor may be a planned evolution or an urgent one. Depending on the prevailing conditions, a submarine may proceed to sea escorted or unescorted. An unescorted departure (single submarine) ensures better concealment, more freedom of movement, and simpler control by the authorities ashore. The main consideration should be concealment of departure, and time of departure. Hence, departure is normally affected by night or during poor visibility, and with total radio silence. The departure may be broken up into three stages:-

- Transit within the limits of one's own Base.
- Transit through any mine-fields or dangerous waters beyond the limits of one's own Base.
- Transit through an established corridor, through area of operations of own surface forces.

In the first two cases, to transit on surface should be the preferred option for the commanding officer to exercise. In the third case he may take the submarine out on surface, or dived. If passing through waters where a hostile submarine or an underwater snooper is probably expected to be lying in wait, the submarine is advised to transit at a depth which will permit her to fire her own weapons, if required. In the case of conventional submarines, the Speed of Advance (SOA) may be so selected as to ensure enough battery power availability after crossing the outer limits of the area where own anti-submarine forces are operating. If detection or opposition by anti-submarine forces of the adversary is expected to be significant, then the submarine may consider sailing out with escorts. Also, when it is not possible to dive and leave harbor, the submarine may consider being escorted out.

Once out of harbor, the command team onboard is advised to ensure that the regime of operation (on surface, dived deep, or at periscope depth) and the speed of advance are in accordance with the mission orders received. If the mission orders do not specify these, then they should be so planned as to safely reach the area of deployment by the expected time and *without compromising her presence at any stage en route*. The speeds of advance and the operating regimes may vary accordingly at different stages. When transiting through areas where the probability of other submarines operating is large, or on detecting an unforeseen or sudden presence of an alien submarine, the commanding officer should consider using silent speeds and transiting at depth.

When transiting through expanses of water where alien submarines are not expected, the commanding officer may consider using mixed regimes for advance. If he encounters extreme bad weather which does not permit him to charge batteries at periscope depth, he may have to resort to surfacing the boat and charging batteries on surface. When evading alien forces, or maneuvering for retaliation in self-defense, he should be restrictive if another one of his colleagues, in another submarine, is also on his way out from harbor (he would have been informed of this before sailing out) and expected to be in the area. If transiting through an area where there is a possibility of mines being present, the commanding officer is advised to proceed at maximum depth while at the same time ensuring that it is at a depth above and clear of the effective range of Ground Mines, depths of water permitting. In areas where multi-dimensional hostile anti-submarine forces are believed to be operating in coordinated operations, the submarine should select a transit depth which gives it relative safety from the most dangerous of those forces. Just before entering the area of deployment, the submarine is advised to ensure that air and battery reserves are maximized.

Once in the area of deployment, it is prudent for the submarine to remain concealed or dived so as not to risk detection. She should get to the area of deployment at the appointed time and on the appointed date. If intelligence about the area and the others units in it is inadequate, it is desirable that a reconnaissance of the complete area be first carried out. All other actions to be carried out, including orders for charging batteries, should be in accordance with the dictates laid down in the mission orders. The mission orders will state the degree and extent of freedom that a commanding officer can exercise in the deployment area.

The return to base port on completion of an assigned task, or because of other reasons (in event of damage or an emergency), is organized by the shore controlling authority. This is normally spelt out to the submarine before she proceeds on her mission. In the event that she has to terminate her mission earlier than envisaged and return to base, special instructions would be signaled to the submarine by the authorities ashore. Such a signal would generally include:-

- The route for transiting back and regimes of operation (speeds of advance and dived, on surface, or at periscope depth).
- Times when required to cross established positions.

- Where to rendezvous whom.
- Intelligence relevant to the area being traversed through.
- Any other information considered necessary.

In wartime conditions or in a hostile environment, **if a rendezvous (R/V) is required to be affected with another vessel**, then care and caution should be observed to ensure that the R/V is accomplished with the right vessel. There are many procedures that can be laid down for the submarine to follow. One such procedure is suggested here. It is only suggestive, and if a better procedure is available, then that should be followed.

The submarine should accurately confirm her position and arrive at the designated R/V area at the appointed time. She is advised to come to the area in a dived state, maintaining a depth that is most advantageous to her underwater detection sensors. If the vessel with whom she is to affect R/V is not locatable, she is advised to periodically come up to periscope depth and switch on her IFF responder (Interrogation Friend or Foe). On locating the vessel in the area she is advised to approach her from her beam, come up to periscope depth, and confirm she is indeed the vessel to affect R/V with. On confirming, she is advised to surface at a distance of 20 to 30 cables from her on her stern and follow further instructions thereafter. Transit to base, after affecting R / V, may be executed independently or under escort. As a rule, the submarine follows the escort and does not precede it. Depending on the situation, the submarine can be escorted dived or on the surface.

In case the R/V is not affected, the submarine is advised to remain in the area and await further instructions from her authorities ashore. In case the submarine comes across another unidentifiable ship in the R / V Area, she is advised to leave the area, taking all precautions to ensure she is not detected, and from a reasonably safe area convey to her authorities ashore about this development after once again confirming the accuracy of her own position and that of the ship. Attacking the unidentified ship is not permitted. However, if another unidentified submarine is detected during hostilities in the area of operations, it must be attacked without any loss of time. If an attack cannot be launched or is unsuccessful, then the submarine is advised to evade the other underwater platform, get out of the area, and from a reasonably safe distance, inform the authorities ashore and the vessel she was to meet, and await further instructions.

SURVIVABILITY CONSIDERATIONS

"The profession of soldiers and sailors has the dignity of danger"

-

Samuel Johnson

The word 'survivability' instead of 'stealth' has been deliberately chosen as part of the heading of this section, and may be looked at in two different ways, in two different lights. The first scenario would involve a deployment where attacking and being attacked are both not imminent, and the submarine is being clandestinely deployed with a peacetime mission in mind. To 'survive' under these conditions would mean not being detected or tracked throughout the sortie, carrying out the mission, and safely getting back, with the opposition being none the wiser of such a mission having ever been undertaken.

The second scenario would involve a deployment in a hostile environment where surviving detection, being tracked, being attacked, and carrying out an attack and getting away, would all be part of the game, and the business of war a reality. Survivability is also a two sided business with those planning the submarine's sortie giving it the right amount of consideration when formulating the mission orders for the submarine, and the latter playing her part to ensure it once she has put out to sea. This section will deal with the actions and aspects to be considered by the planners.

It would be suicidal to throw a submarine into an arena where the odds are loaded heavily against her being able to carry out her mission, and her chances of survival slim. In days gone by, when information gathering was in a rudimentary state and submarines could go in stealthily, gather and relay back much information, there had been instances when compromising and losing a submarine at the cost of gaining vital information through her were found to be acceptable, and therefore resorted to. To lose less than a hundred men and a submarine in exchange for vital information about the enemy that would result in his near total, or total, rout was a very attractive and favorable proposition indeed.

However, in today's day and age, with costly, sophisticated boats in the inventory, and with complicated network centric alternate sources of information available, there is no need to unduly risk a submarine for such an event, and due thought should be given to ensuring her survivability before deploying her. There are a host of minutest details that have to be taken into account to ensure this. Some of the major considerations find mention here, and any or every other factor that may be thought to be important enough to have a bearing on survivability, must also be considered.

It is very difficult to quantify the degree of threat in any situation with some degree of accuracy, no matter how many inputs are available. Environmental considerations, weather predictions, details about the opposition, and what it is likely to do, are at best guesstimates based on available, one-sided, inputs. There, therefore, has to be some way of making a fair guess as to what the opposition is up to. The standard military way of carrying out *an appreciation of the situation* is one of them, which will then lead to various options available and suggest the best course of action to be taken.

The *'theory of probability'* offers another way of arriving at conclusions with some rationality. Using the Theory of Probability and available data and intelligence about the opposition, the Probability of being detected, the Probability of being tracked, and the attacking capabilities of such opposing anti-submarine forces can be worked out.

Let us consider long range airborne elements of the LRMP/MR-ASW variety first. An LRMP/MR-ASW aircraft has extensive endurance and can be deployed at extended ranges of operation from its base. If the opposition possesses such aircraft, they are likely to be encountered while the submarine is on transit to her area of deployment, perhaps in her area of deployment, and maybe even while returning to base, or to any other point for replenishment and turn around. The probability of detection, tracking, and attacking capabilities of these aircraft (with all types of weapons and sensors they can carry) should be worked out separately for the period during transit by the submarine to the deployment area, for the period the submarine is in her deployment area, and for the period that the submarine is expected to move from the deployment area to her next destination for turn around, or in event of damage, for repair. Keeping these figures aside for a moment, let us go to the next step.

Another potent threat to the submarine would be warships with anti-submarine capabilities. They may be with or without anti-submarine helicopters (with dunking sonars, or sonobuoys, or both) onboard. They may have hull mounted or towed sonar arrays. They, like the long-range aircraft, could also be encountered while on transit, in the deployment area, or while coming away from the deployment area. The Probability of Detection, Tracking, and Attacking capabilities of such ships in the inventory of the opposition should next be worked out in a similar manner as has been worked out in respect of aircraft.

There could be underwater threats too in the form of mines or SSN/SSK submarines. If the opposition has the capability to lay mines or possesses SSN/SSK submarines, then the Probability of encountering mines or SSN/SSK submarines while on transit, in the deployment area, and on the route to be taken on completion of the deployment should be worked out, as also the SSN/SSK submarine's detecting, tracking, and attacking capabilities.

Having worked these out separately, the probability of all of them together, or some of them in various permutations and combinations being encountered at a time must next be worked out. To finally arrive at the survivability chances of the submarine, the Measure of Effectiveness (MOE) of evasion from detection by the submarine from these platforms should next be calculated. To arrive at reasonably accurate assessments, it is necessary to have performance details and parameters of the platforms in the opposition's inventory including endurance figures, knowledge of the sensors and weapons they carry, their capabilities, and the parameters of the mines they hold in their inventory, as realistically as possible. This cannot be done overnight. These figures must be data banked during peacetime, constantly updated, and retrieved for the purpose of these calculations when required.

Finally, our own submarine's characteristics and performance details must also be known, and these vary from submarine to submarine depending on the types of weapons onboard, the efficiency of the sensors carried, the age of the submarine, the material state (when the last refit was carried out and when the next one is due), the operational radius assigned to that particular submarine, and depths she can dive to, etc. The survivability figures arrived at will help the planners with yet another input to arrive at the route the submarine should take, the areas where diversionary routes should be looked at, the degree of survivability that can be expected in the deployment area assigned, and her optimum return route on completion of mission. The level of risks decided upon, or found to be acceptable, in order for her to survive will evolve. These figures will also suggest whether and where there is a need to provide direct or indirect support to the submarine, and where other forces may be required to be brought in to neutralize the opposition's effectiveness in those areas, for the durations required (when the submarine is in the area or transiting through).

This kind of attention to detail should be paid even when planning peace time sorties, as a matter of habit. There are many books on Operations Research and Analysis available that will help with algorithms to work out the requirements for aspects that need attention, to ensure planning of submarine operations with optimum survivability considerations.

EVASION AT SEA

"Covered with a rubber like polymer that slicked the surface to add silence and more speed underwater, she had a dark, wet sheen as opposed to the dry look of paint on metal"

-

The Kursk

A submarine has certain emitting physical properties that are dead giveaways, and these can be identified by an adversary to confirm her presence in the area. The most common of these giveaways are self and radiating noises, reflectivity of the hull, electromagnetic field, magnetic field, temperature alterations of the waters around her by her very presence, seismic wave alterations, cosmic wave alterations, radio transmissions, and the likes. There are over thirty such giveaway fields. However, to conceal the submarine's presence it is not necessary to try and suppress all of these tell-tale sources. What need be concealed are only those properties that the adversary is capable of detecting. Reducing emissions of detectable properties can be done to some extent by enhancing concealment measures and ensuring high defensive measures against the adversary through other means.

The main giveaway properties of a submarine are noise, other radiations, and her tracks. These may be interpreted at sea as:-

- Noise – Hydro-acoustic through propellers, and the sound of main and auxiliary machinery running (may be termed as 'primary' noises).
- Radiations – Magnetic field, radio-active field, radio transmissions, 'secondary' noises, electro-magnetic field (e.g. through radar transmissions), diesel engines when snorkeling in transit or charging batteries, wake on the surface, wakes produced by retractable masts, phosphorescence in the wake, etc.
- Tracks - Oil slicks, air bubbles, smoke from diesel exhaust, temperature variations in the water, physical outline noticeable from the air in clear waters when dived shallow, effluents pumped out.

Naturally, concealment of all these 'giveaways' will make the job of the adversary that much more difficult. Decreasing much of the chances of detection of the submarine by those means available with an adversary can be achieved, for example, by reducing time spent on the surface or at periscope depth. This does not mean that, without consulting the hydrological conditions, a submarine at depth is always safe. Even deep, there is an optimum depth where safety is relatively 'high' and a depth where she can still be detected at great ranges.

Hydrology is extremely instructive for a submarine commanding officer and gives him the following information:

- The optimum depth to operate in for minimum noise emission.
- The maximum silent speed that can be ordered at that depth.
- The best depth at which the submarine can do high speeds with minimum emissions of noise.
- Optimum depth to be in to get maximum performance from underwater sensors.
- The best depth to maintain to degenerate the adversary's detection capabilities.

The submarine must, therefore, frequently get all information on hydrological conditions from periscope depth to the *Maximum Operating Depth* in the area, and analyze it carefully. After analysis, the submarine is advised to select the best depth to be at to ensure concealment, or the depth at which her sensors will perform with maximum effect.

Propeller noise is one of the worst giveaways of a submarine's position when

she is underway. Noise level (cavitation) is related to speed and depth of operation. The speed at which 'cavitation' commences may be defined as the Critical Speed. With increase in depth the Critical Speed is higher. *Maximum Silent Speed* is a little less than the Critical Speed. *Maximum Silent Speed* is defined as that speed at which the hydrodynamic noise and the noise of the propeller do not exceed the noise from auxiliaries. *Maximum concealment* is ensured by a submarine when she proceeds dived at great depths (excluding the sound channel depth) at silent speed.

There are various ways of overcoming or evading adversaries. While 'concealment' is one way, the use of 'force' is another. Some of the most successful ways are by operating concealed and singly, by using weapons during hostilities when evasion is difficult, by using the help of own (other) forces to draw the adversary away, by operating near Straits & Narrows and 'choke points' with broad fronts, by being forewarned about the adversary in the area of operations by timely and accurate information/intelligence from the operating authority ashore, by using camouflage and concealment measures to draw the adversary elsewhere, by using Electronic Countermeasures to degrade the performance of the adversary's electronic equipment, by escorting the submarine where required (recall Soviet efforts to get through the GIUK gap to enter the Atlantic Ocean), and by creating openings in anti-submarine booms and nets for own submarine to pass through.

The art of not getting caught harasses and frustrates an adversary, and often brings to bear a disproportionate part of his forces to hunt a submarine down at the cost of his operations elsewhere. The art begins with *'evasion'*. 'Evasion' by a submarine can be split into two types – *evasion from detection*, and *evasion from attacks*. It can be further split into evasion from anti-submarine surface forces (ships), anti-submarine forces from the air (fixed wing and rotary wing), and anti-submarine submarines underwater (SSNs/SSKs). Let us briefly examine the main features of the art of evasion from these different adversaries just enumerated.

Evasion from detection by surface ships of the adversary operating in a given area could be achieved by avoiding or sidestepping that area altogether. However, if avoiding or sidestepping is not possible, then the submarine is faced with a situation wherein she has to discreetly force her way through, and come out of that area without being detected.
In another scenario, she may be faced with a situation where she has to not only discreetly pierce through a cordon of searching surface ships, but face the added worry of possible static seabed sensors in the vicinity. The first option would always be to side step. Getting out of the search area of hostile forces involves side-stepping from their paths, using all discreet sensors for detection and tracking at the disposal of the submarine so as to carry out the maneuver in time i.e. before the submarine is itself within the opposition's range of detection. Side-stepping is only possible when the submarine has a detection range advantage. When a range advantage is not available, it is recommended that the submarine opt for penetrating the formation screen. Before outbreak of hostilities, this is done discreetly. During hostilities, this is best done by carrying out an

attack and forcing one's way through in the ensuing melee.

Evasion from attack by the adversary's ships is to be handled slightly differently. Hostile ASW forces will use any or all of the anti-submarine weapons at their disposal to attack a submarine. To evade an attack, a combination of alterations in course, speed, and depth are recommended. When maneuvering to evade a tactical nuclear attack, the submarine should endeavor to put maximum distance between her and point of aim (which, in actual fact, will be the present position of the submarine.), and to minimize the effective area alter course, speed, and select a depth beyond which the reflected shock waves from the surface will not lethally buffet the submarine again and again. Care must be taken to ensure that the submarine does not go too deep where the radius of effectiveness of the bomb increases rapidly. The maneuver to evade homing torpedoes with conventional warheads is based on the principle of quickly getting away from the point of aim, and reducing the effectiveness of the Homing Head which involves a radical alteration of course and depth, and an increase of speed based on what the best recommended speed is as per hydrology in the area.

When attempting to *overcome anti-submarine fixed wing aircraft*, it is prudent to remember that air ASW effort can be brought to bear on a submarine anywhere within the range of operation of the aircraft. At the same time, it must be remembered that its time on task will progressively increase from its furthest point of deployment as the deployment area gets nearer and nearer its operating base. The effectiveness of air ASW effort will depend on weather and meteorological conditions to quite an extent. The capabilities and limitations of airborne platforms and their sensors affect their ASW performance too. As a rule, it is recommended that a submarine operate deep in areas where Air ASW forces are expected to operate. Operating at Periscope Depth should be restricted to **short** durations required for charging batteries, topping up air, or because of other unavoidable reasons. Such durations should preferably be in dark hours or during periods of poor visibility. Every aircraft detected should be considered to be hostile, and the submarine should take evasive measures, unless an R / V (rendezvous) has been scheduled with one of them. Evasion includes diving deep and getting away from the spot from where the aircraft was detected before the latter detects the submarine, and keeping clear of the attacking zone if it is felt that the aircraft may have suspected the presence of the submarine.

It would be prudent for the submarine to assume that the aircraft has dropped some sensors in the water to localize the submarine's position. It is therefore recommended that the endurance of such sensors and the endurance of the aircraft be considered while maneuvering underwater, and before an attempt is made to return to periscope depth. If, for whatever reason, the submarine is not able to dive deep (damage, shallow waters etc) she is advised to maneuver to evade the dropped weapon with sharp course alterations from the moment the aircraft has steadied up on its weapon launching course, and drastically increase or decrease speed the moment the weapon is dropped. If a torpedo has been dropped, the submarine is advised to alter courses based on laid down doctrines on the subject. *Evasion from rotary wing ASW aircraft* can also be split into evasion from detection, evasion from tracking, and evasion from attack. Tactical doctrines should be formulated on how to evade helicopters with dunking sonar, and those with sonobuoys, and these should be adhered to.

The most difficult platform to evade is another submarine of the hunter-killer variety, referred to as the SSN or the SSK. Both having the freedom of use of the vertical plane; the reaction times after detection are far less than in respect of surface ships or aircraft. Having said that, for a submarine to hunt another submarine in the wide open ocean is like looking for the proverbial needle in the haystack The SSN or the SSK would be deployed carefully, and in *a 'high' to 'very high' submarine probability area.* The SSN or the SSK would have the ability to seek and destroy both nuclear propelled and conventionally propelled submarines independently, or with the assistance of aircraft, ships, or static seabed sensors. Evasion from such a platform would include carrying out an urgent attack and destroying her, avoiding the area where she is expected to be operating, silently slipping past her through the area where she is expected to be operating in, or avoiding detection and tracking by her.

Given that the SSN/SSK will have a detection advantage under normal circumstances, and so as not to be taken unawares, it is necessary when entering areas where the adversary's SSN/SSK submarine is expected to be patrolling, to use silent speeds, select a depth that will give the advantage of concealment, and select a course to get out of her path as early as possible.

Similar options exist for evading being tracked, and being attacked. Therefore, tactical doctrines should lay down similar procedures to counter the SSN/SSK, and to avoid using changes in course, speed, and depth that are dead giveaways. During hostilities both have the option of attacking each other and it is best to be the first one to carry out an attack.

The skill of not getting caught should be repeatedly practiced during work up and developed into a fine art against all types of platforms till it becomes second nature to the submarine. It enhances the concealment abilities of the submarine and enables her to get on to fulfilling the main mission assigned to her.

INTELLIGENCE RECONNAISSANCE

If done correctly, that it was done at all would never be known!

- *Anon*

Conventional submarines are sluggish and have a low speed of advance when compared to nuclear submarines, ships or aircraft. But they also, like nuclear submarines, have the advantage of operating concealed by using the third dimension below the surface of the sea, which is not usable by the other two platforms. Once discovered, however, the submarine becomes defensive and does all she can to get away and regain the initiative. If the mission she is on is of a clandestine nature, it could get compromised. However, if there is a requirement for a platform to discreetly go into sensitive waters and bring back information, a submarine is the ideal choice, provided a possible delay in that information getting back is acceptable. If she is required to gather information and communicate it from her area of deployment or near it, she could be endangered, and her survivability could be in peril if her communication channels are not secure.

This must not be taken to imply that she should not be tasked with risking her presence, and be directed to communicate information. When *the information required is so vital that it reaching the Command ashore or the Battle Group at sea, at the earliest, is more important than risking / losing a submarine, she may be asked to be indiscreet and relay that information from her area of deployment, or near it, using one of the safer means of communications available. This should be the rarest of rare instances.*

As a platform, she is good for reconnaissance under certain conditions. A submarine can be deployed at great ranges where surface and air reconnaissance forces cannot be deployed. She can be deployed for great lengths of time to keep watch over the adversary in distant areas. She can also be deployed effectively, irrespective of weather conditions. She can bring back documentary evidence by way of photographs and recorded sounds. She can land personnel in areas they are not supposed to be in, to gather information or perform an act they are not supposed to perform in this manner, and bring them back safely, with the opposition being none the wiser.

When deploying submarines for reconnaissance, it must be borne in mind that some of its sensors will be handicapped with limited ranges of detection and as a consequence limit the reconnaissance effort. Two-way communications may not always be possible when the submarine is below Periscope Depth. Possible delays in receiving information from the submarine should be acceptable. If a conventional boat is being deployed, then her need for periodically charging her batteries must be taken into account in the planning stage.

Occasions that would weigh in favor of deploying submarines for intelligence reconnaissance would include conditions when information on the nature and characteristics of a particular object to be reported upon is required, or the military-geographical conditions in a particular area are required to be known.

A submarine is the better choice when harbor defenses and pattern of deployment of local naval defenses of the adversary need to be known, provided the depths of water in the area is adequate. Sometimes a requirement comes up to discern the type of camouflaging and deception that is being adopted by the adversary, or the physical features of a particular object or area are required to be known. A submarine could be the obvious choice to find such information. The adversary may use radars and other electronic emitting services in certain patterns. If the pattern is required to be known, deploying a submarine for gathering such information would be most useful. When hydrological and weather conditions in a particular area preclude the use of other platforms for reconnaissance, a submarine would be the preferred choice. Finally, when stiff opposition is offered by the adversary in the region, a submarine would once again be the choice.

From all that has just been stated, it follows that there are a number of tasks assignable to a submarine, literally under the adversary's nose, and these include getting to understand the pattern of harbor patrols, the pattern of turnaround for replenishment of the adversary's local naval defense units or anti-submarine forces, the communication channels he is using; locating, from the pattern of operations of covering forces, his swept channels or routes taken by his ships and submarines, and their marshalling areas (if assigned); establishing their methods of leaving harbor, the way they form up outside harbor, the composition of forces in harbor, the corridors used by their aircraft going out and coming in for landing (if an airfield or helipad exists in or near the harbor); establishing the types of defenses ashore and their command and control support; recording Electronic Warfare (EW) inputs by way of radio and radar frequencies being used by merchantmen, warships, shore units, and aircraft; recording sonar information of hostile units; studying the navigational marks, buoys, lights, hydrological and other conditions in the area; and gather any other information of this type that is required..

It may not always be possible for nuclear submarines to go so close as to gather some of the information listed above, as they are more comfortable outside the 30 meters sounding line. Conventional submarines are more suited to go inside the 30 meters sounding line to gather such data. In the open seas, however, either of them can be deployed to study, gather, and in some cases report the composition of forces, the type of screening, the main body composition, the number of fixed / rotary wing aircraft in the area / with the forces, their course, speed, and time of passing through the area; the frequency, type and nationalities of shipping transiting through a particular area; the proximity of the adversary's strike force or battle group, their composition, and their PCS (position, course, and speed); any changes in the adversary's pattern of operations, or communication channels in use, or composition of forces (splitting up, augmentation, etc) in the area; anything unusual observed in the area of deployment; and anything else required to be conveyed that is not listed above.

Other than the methods adopted for a search by the submarine and its sequence of actions, generally speaking, reconnaissance depends on the type of mission assigned, the geo-physical conditions in the assigned area of operations, the capabilities and limitations of the sensors being used by the submarine, and the tactics being employed by her. Some of the main methods considered by planners ashore are:-

- To deploy the submarine in a limited, bounded area, and carry out reconnaissance within that area.
- To send the submarine looking for the adversary in a specific area.
- To track a moving adversary who has already been discovered earlier.

Reconnaissance in a finite, bounded, area is the method

recommended when the mission is to be carried out outside the adversary's base, port, or coastline. The positive aspects of this method of deployment is that the submarine's position is always known to the supporting staff ashore with reasonable accuracy, and she can be tasked with relaying information constantly about hostile forces, or the situation and changes, as they exist, off that area, with the use of her own sensors. In the case of conventional submarines, the requirements for battery charging should also be catered for.

Reconnaissance in the open seas by a submarine is planned to locate the adversary with the aim of bringing weapons to bear on him or with the aim of homing other supporting forces onto him. An SSN would be the preferred platform but that does not preclude the use of a conventional submarine. The area selected is finite when it is reasonably certain that the adversary is expected to pass through that area, or when he is already located there. Reconnaissance may be organized along a barrier line – or across it – when the adversary is almost certain to pass or intersect it. Such a possibility may occur in restricted waters or at the entrances of Straits and Narrows. A submarine is sent looking for the adversary in a given direction or particular area when:-

- The route taken by the adversary is generally already known.
- With the help of an earlier reconnaissance, the place and direction of movement of hostile forces has already been discovered.
- The carrying out of a given mission results in detection of the adversary in a particular direction at an early stage.
- There is a need to sanitize a particular area or lane or channel, through which our own submarines may be transiting, and which is suspected of being patrolled by hostile submarine/s.

On detecting hostile forces either through their electronic emissions, sonar transmissions, or by any other means, the submarine has one of two choices. If the electronic transmissions are strong, and from an aircraft, then immaterial of the situation, she is advised to take immediate evasive action by going deep and maneuvering underwater as part of aircraft evasion tactics. If the transmissions are from an aircraft, but weak, it may be prudent to remain at Periscope Depth, determine all the characteristics of the radar, and direction of movement of the aircraft. If the electronic emissions are from a surface ship, then the submarine is advised to close the ship to maximum detection range of the submarine's passive sonar. Once the bearing and bearing movement of the ship is determined, the submarine is advised to work out the ship's course, speed, and range, and then close her to detect the others with her, and the formation they are in. Thereafter, if the orders are to track the formation, the submarine is advised to take up a parallel course outside the detection range of the adversary, and track the formation.

Reconnaissance is a peacetime activity that submarines are always called upon to do very frequently. The aim is to gather data for the data bank and to understand the adversary's pattern of operations. No doubt, in a hostile environment there will be some changes in his behavior and ways of operating. However much of what he does in peace will remain unchanged during hostilities, and hence the need for peacetime reconnaissance. It is also a wartime activity that often culminates in an attack with the aim of destroying the enemy. The task becomes easier if practiced in peace.

SSK OPERATIONS

The best anti-submarine warfare platform is invariably another submarine.

- Anon

On the 3rd or 4th of February 2009, two modern stealthy 'boomers', one the *HMS Vanguard* of the Royal Navy, and the other *Le Triomphant* of the French Navy, rammed each other underwater in the Atlantic Ocean, with neither aware of the other's prior presence in the area. The official explanation given was that both were so quiet that they could not detect each other (sic)! This occurred, despite both having perhaps the costliest and most sensitive sonar equipment in the world. The *Vanguard* carries 16 Trident missiles with up to 8 warheads, and *Le Triomphant* 16 missiles with up to 6 warheads. They obviously had overlapping patrol areas, and were targeting the same country's cities with missiles of identical range. They may have been operating at silent speeds (with own noise less than the ambient sea noise) and at 'optimum depths'. Both survived and returned to their respective harbors. NATO countries exchange information about the areas and depths at which their submarines operate on patrols. France opted out of NATO and is not bound to divulge such information. Such collisions are rare, but have occurred in the past under different circumstances, and will continue to do so in the future, when two very quiet and stealthy platforms play cat and mouse games..

During the Cold War there were more than twenty incidents of Soviet and Western submarines colliding with each other. Some were intentional and some accidental. Some sank and others limped home in a damaged condition. Most of these incidents were not made public till many years later. Details are now freely available from both sides. With scores of submarines milling around in the oceans of the world, there is a need for water space management in peacetime, at least amongst allies. Patrolling as the third arm of the 'triad' and tracking each other's submarines under seemingly placid seas carry on with regularity even after the end of the Cold War.

SSK submarines and SSNs are probably the best anti-submarine platforms as they operate in the same environment as their victim.

Whatever advantages the victim attempts to take from the environment are negated by these predators operating in similar conditions with superior facilities at their disposal, and in a very quiet and alert manner.

During hostilities, SSK submarines are normally assigned tasks in straits, narrows, choke points, restricted waters, or known submarine deployment areas of the adversary, to seek and destroy their submarines that are potential threats to own forces and the nation because of their massive fire power and destructive capabilities. Unlike SSNs, they do not have the speed or sustaining power to seek and tail adversaries continuously. During peace-time operations they may be assigned the task of monitoring the periodicity of traversing these choke points by SSBNs and SSNs of the adversary with a view to determining their pattern of operations, or plotting routes and transit channels that they are likely to use in the future.

SSK submarines may also be deployed off own harbors to monitor activities of unfriendly submarines during peace, and to destroy them during hostilities. They may be deployed in any area where the presence of an unfriendly underwater visitor may be expected, on similar missions.

Similarly, SSNs have the challenging task of seeking and tailing SSBNs during peacetime with the responsibility of blowing them up when the balloon goes up and before they release their weapons of mass destruction. The peacetime task of trailing SSBNs is a very demanding one. It is not continuous but a process of gaining contact, holding contact, losing contact, regaining contact, and so on. Once the SSBN gets suspicious of the SSN's presence and goes 'silent' in an attempt to shake her off, the process becomes even more tedious. Then there is the possibility that the SSBN is operating with her own SSN covering her, in which case trailing the SSBN would entail endeavoring to constantly hold her on sonar while avoiding the opposing SSN. She would probably attempt to stay on the opposite side of the escorting SSN, keeping the SSBN between them all the while.

For anti-submarine submarine operations between nuclear boats, the task has been made somewhat easier by the very fact that bastions and deployment areas of one is known to the other with a fair degree of accuracy. This is because the efforts put in to discover each other's submarine probable areas over the last fifty years or so have been very intense. In the case of conventional submarines hunting each other, the efforts put in to locate deployment areas have not been so intense. As a result, the task of trailing during peace-time, and attacking and destroying another submarine during hostilities is not an easy one and successes have been very limited in the past. A look at the Tables below that give data of submarines sunk during WWII will make this obvious. Some of these were chance encounters. (Here it must be remembered that specialized SSK submarines were built only after World War II).

	No of Submarines built			**Number of Submarines Sunk**			
Country	*Up to WWII*	*During WWII*	*Total*	*By SU Forces*	*%*	***By SSK's***	*%*
Germany	57	1131	1188	781	63	**23**	2
Italy	115	41	156	84	54	**20**	13
Japan	63	129	192	130	66	**25**	13
UK	58 (69)	165	223	76	34	**5**	2
USA	112	203	315	52	17	**2**	1
Total	405 (416)	1669	2085	1123	54	**75**	4

SSK submarines are designed and built for silent, autonomous, operations for longer durations in the area of deployment than other conventional submarines. They are capable of operating independent of the prevailing weather conditions in the area, and at optimum and favorable depths from where they can seek and destroy, or track, or evade other conventional submarines, SSNs, and SSBNs under conditions favorable to them. They are adept in operating in shallow waters. They can also be assigned tasks that are assigned to other ocean going submarines against shipping, as primary or secondary tasks. In a similar manner, while SSNs are designed to carry out hunter-killer functions against SSBNs and/or SSKs, they have a multi-tasking capability and enjoy the flexibility of being deployed on varied roles. Despite the special considerations that go into designing SSK submarines, they do experience difficulties in classifying detected submarines due to low noise emitted by the latter. They are therefore, generally deployed singly so that any submarine contact they gain can be presumed to be hostile and classified as such. The contact can then be tracked/attacked without the need for the SSK going 'vocal' with IFF, underwater communications, and the likes.

When planning SSK operations, the area and operating depths must be carefully selected so as to give them a distinct advantage over the adversary. Of course, it is possible that the adversary also selects the same area and depth with the same intentions. Under the same hydrological conditions, it is therefore possible that they may simultaneously detect each other. In most cases, because of the silent nature of these platforms, detection is invariably at reduced ranges when weapons can be immediately fired by either the hunter or the hunted. During hostilities the *aim should always be to fire first.*

When the staff ashore decides to deploy an SSK submarine against the adversary's underwater force, they must select the deployment area with utmost care. It could be in the approaches to the adversary's harbor, or where the adversary's submarines are known (intelligence) to operate, or, where they are expected to charge their batteries. She could also be deployed along routes that the adversary's submarine/s are expected to pass, especially in the vicinity of choke points.

RE-POSITIONING SSKs BASED ON DATA RECEIVED FROM STATIC SEA-BED SENSORS OF PRESENCE OF HOSTILE SUBMARINE.

In open seas, SSNs could be deployed within Limiting Lines of Approach (LLAs) of own surface forces, and ahead of them, in clearly defined 'kill zones' in which only they are allowed to operate and shoot. The aim would be to attack and sink a waiting SSGN or Fleet Attack conventional submarine of the adversary before the latter launches its missiles or carries out a torpedo attack on the main body being screened – most likely an aircraft carrier. The surface force would rely on mobility for its defense, and when high speeds are used by them, so must the SSGN or the conventional boat so as to get into position to attack the force. Speed reveals any submarine's vulnerability. It creates noise and degrades its sonar's listening capability. The escorting SSNs positioned ahead of the surface forces will know the force speed and the LLA for the SSGN or SSK at that speed. They can position themselves for an ambush. If any of the

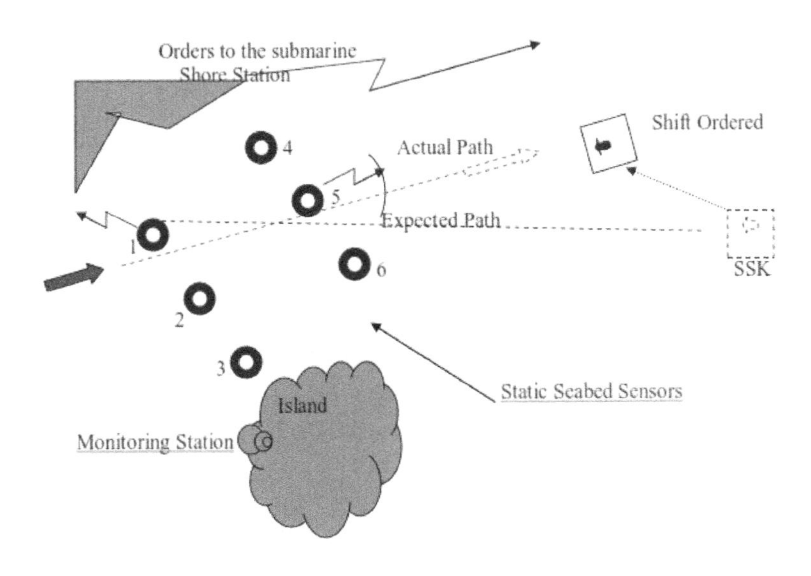

Orders to the submarine
Shore Station
Shift Ordered
4
Actual Path
5
Expected Path
1
6
SSK
2
3
Static Seabed Sensors
Island
Monitoring Station

INDIAN NAVY'S SHISHUMAR CLASS SSK

surface forces are towing a Surveillance Towed Array System (SURTASS) she can forewarn the surface force and the SSNs of an adversary in their path or area. The SSNs would typically operate in the sprint and drift mode – alternately race forward and then slow down to listen. The aim would be to detect the adversary first and launch torpedoes before he launches his attack.

For conventional submarines on anti-submarine missions, it is prudent to plan all movements and maneuvers at *minimum* silent speed in their quietest operating regime. The selected speed should give the SSK maximum search productivity. This is necessary for them to gain the advantage of initial detection. Alteration of speed and regime of operation is not recommended once the adversary is detected.

When selecting the operating depth in the deployment area, it is necessary to study the bathythermograph profile in the area, geographical restrictions if any, depth constraints on the use of her anti-submarine torpedoes, and the probable operating depth of the adversary. If refractive restrictions are absent, the recommended depth would be the operating depth of the SSK. If a layer is present, then the selected depth would be on the opposite side to that (above/below) of where the adversary's Passive sonar is located. If a sound channel is present, it is recommended that the SSK operates close to the axis of the sound channel. In all conditions, the thumb rule would be to select a depth that gives the SSK maximum detection advantage over the adversary. The best position for an SSK submarine to be in while tracking another submarine is in his stern sector.

The SSN versus SSK, or vice versa, situation is a very interesting and intriguing operation. The strengths and weaknesses of each against the other notwithstanding, a little external help to each, can alter the odds against the other. If we suppose that the SSK is going to lie in wait for the SSN at the mouths of Straits, Narrows, or choke points, it would be reasonable to presume that one or the other will be aided by sea bed sensors, and/or airborne platforms in the area. Straits and Narrows are where seawater mixing from one sea to another takes place. They are also likely to present shallow waters in the immediate vicinity. Ambient noises are likely to be higher and passive sonar receptivity, poorer. SSNs, with bottom bounce techniques, can use active sonar and get ranges of SSKs at distances of 20,000 yards or thereabouts. Their active sonar is a concentrated, directional, beam and not an all-round noise maker. However, when receiving active transmissions from ranges of 20000yards away, the SSK is likely to receive it all round and not be able to make out its direction or range. The SSK can only confirm the presence of the SSN. On the other hand, static sea bed sensors (like the SOSUS chain in the Greenland – Iceland – UK chain) placed properly can alert an SSK of the SSN's whereabouts sufficiently early for the SSK to react. Anti-submarine fixed wing aircraft can also drop sonobuoy sensors and alert one or the other of the adversary's presence and actions. Either or both of them can benefit by coordinated operations with other platforms in the area. It is a crafty cat and mouse game with one certain outcome – one or the other is not going to return home.

TORPEDO ATTACKS

"It's a hard fight with a short stick"

- A Submariner

In today's context, a torpedo attack by a submarine on a surface combatant unit would be considered to be a close encounter, what with beyond-the-horizon ranged tactical missiles and cruise missiles in the inventory of modern boats. Most modern torpedoes also have beyond the horizon ranges. One torpedo – the Russian *Shkval* – even has comparative speeds. Even so, the temptation would nearly always be to launch a missile from extreme, effective, ranges after 'bracketing' and properly classifying the target, and quietly slip away. Yet, there will be occasions when a submarine would have to engage a target with torpedoes, and this form of attack brings with it all the ingredients of an exciting, dauntless, challenging, and cheeky assault on the adversary.

For a submarine to carry out a successful attack, she needs to first *'classify'* the contact and confirm that it is hostile and attackable before going in for an attack. This is very important in today's context where there are many platforms of different nationalities milling around in the same waters, each of them looking after their own self interests. Merchantmen chance their arm in troubled waters to get to their destinations earlier, coalition forces home on to troubled spots (and they are the last ones a submarine would want to sink unless they form a part of the 'other' side), warships of countries not directly involved in any sort of skirmish go about their business, fishing boats pursue their search for a daily income, and so on. *Identification and classification* are of the utmost importance before an attack is carried out. Any or all her onboard sensors may be required to do this – her low frequency passive sonar, passive medium range low frequency towed array, very low frequency long range towed array, the active sonar interceptor, and whatever other relevant sensors she has onboard.

Having confirmed that the selected ship is the target, let us now examine what a submarine commanding officer does to carry out a torpedo attack. We are now looking at a very basic attack on a single target. To begin with, the attacking commanding officer must know the target's relative or geographical position with respect to his own position. He must next discern which way, and at what speed, the target is moving with respect to his own submarine's position. If she is opening out with a speed advantage, she may not be attackable with torpedoes. If she is closing in, the submarine still must eventually be able to get to within attackable range for torpedoes to be fired at her. 'Doppler' will assist him in determining this, and his sonar operator will discern it quickly and report to him.

THE BASIC TORPEDO TRIANGLE

To get into an attackable position, the commanding officer must position his submarine within the Limiting Lines of Approach (LLAs) of the closing target. He may have to just lie in wait, and wait for the victim to come his way, or edge slowly into her direct path and wait for her to close, or move at speed to get into the LLAs. Target 'range' is required to assess whether the target is reachable. 'Target Motion Parameters (Course, Speed, and Angle on the Target's Bow)' are required to determine the position of the target at the moment of establishing contact, and at the moment of releasing torpedoes.

From the moment the commanding officer detects the target through any of the submarine's sensors, these parameters are worked out by the Fire Control Computer (FCC) of the submarine, backed up by manual plotting in the chart house or operations room, well before the submarine is maneuvered to get into torpedo firing position. There may be other ships in the vicinity, and this makes the selection of target that much more difficult. While any one of the submarine's *active sensors* would give the target's initial bearing as well as the range, her *passive sensors* may only give her a somewhat reliable target bearing and rate of change of bearing, and, perhaps, a not too reliable range, presuming that the initial detection, identification, and classification has been carried out at some distance away, and well beyond the torpedo's firing range. In an enemy probable area, while going in for a torpedo attack, it is more likely than not that the attacking commanding officer will only have his submarine's passive detection sensors in operation, and all they are going to give him is a decent initial bearing and bearing movement rate of the target from the submarine's own position. The FCC and the manual plotters register these readings and get down to work.

From the first few bearing inputs from the submarine's passive sensor, the FCC would work out the direction and change of target bearing movement with respect to time. To achieve this quickly, the simplest method would be by nullifying the effects of the submarine's own movement with respect to the target, with the resultant effect of all changes being attributed to the target alone. This the commanding officer does by pointing the submarine towards the target and reducing speed to near 'zero' knots (almost stopped). The rate of change of the target's bearing will then be attributable solely to the target's movements.

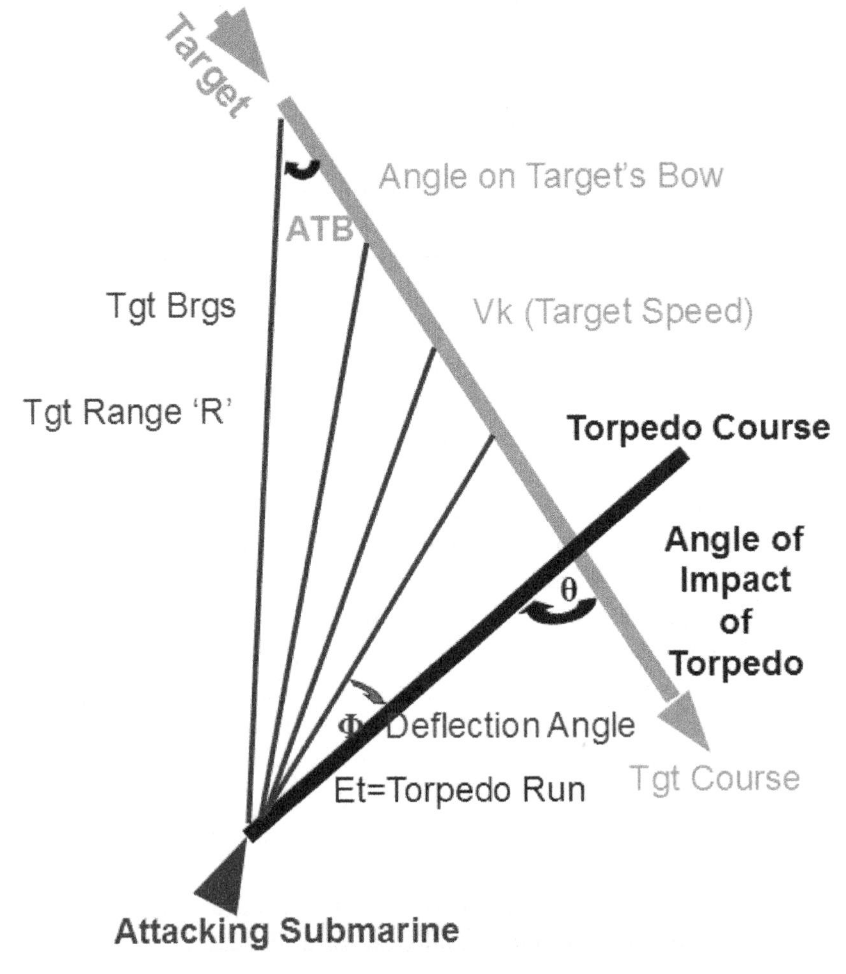

Fig - THE CLASSIC TORPEDO TRIANGLE

The commanding officer knows that it is logical to expect a minimal, barely discernible, rate of change of bearing when the target is at some distance, as opposed to when she is relatively closer. If the range of the target is not accurately discernible through passive detection sensors (especially at extreme ranges of detection of the submarine's sensors), or any of the other methods available (bottom bounce, lower frequencies of target radiation, periscope ranging etc), it would be prudent on the part of the commanding officer to assume a certain 'expected target range' initially, that may be arrived at after taking the detecting sensor's capabilities and other extraneous but relevant conditions into consideration, and feeding this as an initial input to the FCC and the manual plotters. By observing the target's change in bearings over fixed intervals of time at that initially assumed range away from her, it is possible for the manual plotters and the FCC to work out a rough estimate of the target's course and feed this information to the commanding officer.

Across this period of time that the 'change in bearing' and the 'bearing rate' was being observed, the sonar operator from the sonar room would have been able to make out whether the target was closing the submarine or opening out, through Doppler readings. This information is important to the commanding officer, as bearings can change at an equal rate when the target is closing or opening, on certain courses. The need to eliminate one or the other is necessary.

At this stage, the commanding officer has an approximate target Course at an approximate target Range. There may be slight differences in the figures worked out by the manual plotters and the FCC. The commanding officer selects mean values or rejects one or the other. They may well be the same! In any case, he takes a decision and directs the FCC operator and the manual plotters to set a fixed Course and Range, as decided by him. With these parameters, it is necessary for the manual plotters and the FCC to next work out the 'speed' of the target.

The commanding officer maneuvers his submarine on to a seemingly collision course with the target, which is still some distance away, so as to get a constant, steady, bearing of the target with respect to the attacking submarine. He will also take care to ensure that these maneuvers do not make his submarine 'indiscreet' and compromise her position in the process. The FCC and the manual plotters receive data from the submarine's sensors, gyro, log, etc, on a continuous basis as also any other inputs, corrections and commands that are separately fed to it, and automatically gives a continuous feed of the target's current/running co-ordinates, while progressively increasing the accuracy of these parameters. The FCC simultaneously works out and recommends how best to get away from the target or its aiming point at any time, should a contingency arrive wherein the commanding officer finds a requirement to break off the attack and get away.

If the target had been doing a steady course at a steady speed and closing the submarine all the while, the commanding officer and his attack team would have found the task of determining the target's parameters quite easy. All targets do not necessarily move obligingly on a steady course at a steady speed in a 'submarine probable area'. They would be weaving or zigzagging, or superimposing one on the other and doing both constantly to deny a lurking submarine the opportunity to accurately find or work out the parameters that she requires before firing her torpedo at the target. The FCC onboard the submarine is fairly quick to alert the commanding officer about these alterations of courses and changes in speeds of the target so that he can re-maneuver and rework target parameters each time the target alters course or speed.

There are a number of other factors that can complicate matters for the attacking submarine. The submarine's method of detection, classification, and tracking of the target is mainly with the help of 'sonar' or sound, assisted by the FCC that does all the computation work, and the manual plotter. A few examples will elaborate this. Sound travels at a certain speed in water. It therefore takes a certain amount of time to travel and reach the submarine from the target. During the time that it takes to reach the submarine, the target would have moved away from that spot from where the sound emanated, by a certain displacement that would depend on the course she was steering, and the speed at which she was moving across the water and away from that detection spot. The parameters that reach the submarine from the target are already made inaccurate by this displacement. While detecting and classifying the target's active sonar transmissions (which are detected much before the target gets to detect the submarine) the target range can be worked out using her passive ranging sensors. It will not be accurate at extreme distances and will have errors. There are time lag errors, instrumentation errors, target discrimination errors in terms of bearing and range errors at great ranges (if more than one ship is in the vicinity), etc. Underwater IFF will have errors. Next, owing to varying bathy thermograph conditions of the medium through which sound travels (i.e. temperature, salinity, and pressure of sea water), the sound need not travel in a straight line from the target to the submarine. This would lead to further inaccuracies in the bearing of the target, as read, and the range of the target as calculated. The FCC and the manual plotters attempt to correct or negate all these errors to present the best possible corrected picture to the commanding officer. In the process of eliminating errors, the commanding officer may have to put the submarine through certain additional maneuvers.

Having finally worked out the target motion parameters as accurately as possible, the commanding officer will next want the FCC and his chart house/operations room team to work out the well-known *torpedo firing triangle* which has to be solved before firing a torpedo. Even with sophisticated homing torpedoes onboard that may be fired at the target's wake, or on a target's bearing and thereafter left to seek and attack the target independently, or led to the target through the trailing umbilical data transmission wire if it is wire-guided, the *torpedo firing triangle* must be solved. When worked out, the torpedo triangle gives the bearing of the target, target course, target speed, target range, the 'angle on the target's bow', the firing course the submarine must get on to, the deflection angle for torpedo firing, the torpedo course, the torpedo speed, the torpedo running distance, and the angle of impact of the torpedo on the target for a straight running torpedo. The FCC will next feed necessary firing data to the torpedo tube systems and the torpedo's control systems. It will work out the point of aim.

The point of aim need not always be the center of the target. For instance, on hearing torpedo Hydrophone Effect (HE), a target may take evasive measures. In that case it will not maintain a steady course and speed. It will violently alter course and/or speed (usually decrease). So an area has to be defined within which the target will be situated even after beginning evasive measures. The point of aim could be the center of this area. One can argue that with modern day torpedoes, this kind of information is not required. The counter argument to that is that the FCC, by working all this out, gives the option to fire a non-homing torpedo at the target, or one with its homing devices switched off, to neutralize countermeasures thrown in the incoming torpedo's path. It may be recollected that despite carrying modern torpedoes onboard, HMS *Conqueror* opted to fire a straight running torpedo of WWII vintage at the Argentinean cruiser, *Belgrano*, during the Falklands conflict of 1982, with telling effect. A straight runner with an inert homing head is immune to all electronic and sonar countermeasure deployed by a ship. Some torpedoes aim for the propellers while others get on to the target's wake and then catch up with it. Different types of torpedoes, with different types of homing heads have different requirements for where to aim. Their requirements are to be satisfied to achieve a hit.

THE EFFECT OF A SINGLE TORPEDO HIT

The commanding officer then chooses his combination of torpedoes and fires. For him and the submarine, having compromised their position with the release of torpedoes, the process of getting away and regaining the initiative next begins. He may well have to turn around and attack another target in the vicinity, for which the FCC would have worked out details. Getting away amidst all that confusion after direct hits gives the submarine the best chance for escape.

All through the attack, the FCC and the manual plot were presenting a visual picture of the overall situation to the commanding officer to assist him in his decision making. The sound ray path plotter assisted him in arriving at the best firing depth from a study of hydrological and other conditions. In case the target was protected by screening ships, the FCC and manual plot would have also worked out and suggested the best options for screen penetration, evasion etc., under prevailing conditions. After firing torpedoes, the FCC is also capable of analyzing the success or failure of the attack. Additionally, the FCC provides training value to the attack team during peacetime.

What has just been described is an undisturbed attack on a single target with an archaic system supported by an analog computer to understand the very basics of torpedo firing. Modern day requirements are more complex, and there is need to track a number of targets simultaneously, and even attack more than one target at a time. Submarines are no longer restricted to carrying only torpedoes onboard. Modern boats have mixed weapon compositions of torpedo tube launched missiles, separate dedicated missile launchers, dedicated anti-shipping torpedoes, dedicated anti-submarine torpedoes, mines, and so on. A contemporary submarine at her best can spit out missiles *and* torpedoes effectively and with venom, be it an SSBM, an SSN, or a conventional boat. Inputs to the computers must now come from great ranges, especially for missiles. These inputs may come from other platforms ahead of the submarine to extend her 'eyes' and 'ears". More often than not, they do.

As stated earlier, FCC's now have to be capable of handling much more than a single-ship attack. We will briefly look at a modern FCC to see where technology has taken us today. There are different FCCs onboard different modern submarines. They have all got some common features, and at the same time, each of them has exclusive features that set them apart from one another. The modern FCC accepts target coordinates in terms of relative bearings, as discerned by the submarine, or passed on by an airborne platform or any other unit. It can also accept geographical coordinates. It can track and follow more than one target simultaneously, even while concentrating on attacking one target at a time. It gathers particulars and presents the state of weapons to the commanding officer. Hydrological conditions are also presented by the FCC to the commanding officer. Depending on its level of sophistication, the FCC presents relevant information about the tactical scene to the commanding officer. All these inputs are generally given as a function of 'time' and are therefore guiding influences for the computer. That is why the accuracy of the final solution, in most cases, depends on the errors in initial inputs, and the laws of change that influence these initial inputs with the passage of time.

In earlier wars, submarines endeavored to penetrate screening ships placed round a vital target, to get to the latter and attempt to sink it. The protected target could have been an unarmed merchantman carrying precious cargo, or the very sparsely armed fleet replenishment tanker, or even an aircraft carrier. In modern day wars, getting even a screening ship or two would have an equally devastating effect on the adversary. The choice of target may either be stated in the mission orders received by the submarine, or left to the submarine commander to decide. One torpedo hit on an escort creates pandemonium in a formation and advantage could be taken of the ensuing melee to sink a few more hapless ships, if such a situation comes one's way. Such opportunities should not be lost as they rarely present themselves in undersea warfare today.

New generation submarine tactical data management systems use current computer technology to handle more than one contact at a time. Some can handle at least 350 to 400 tracks simultaneously, maintaining up to 6 to 8 hours of history of these tracks. There are others that can handle 300 sensor tracks for up to 6 to 8 hours or 100 tracks for 24 hours, or even 20 targets and 3 to 4 different weapon systems at a time! One such system employs a client/server or netted architecture. Data from sonar and navigational systems are forwarded to a server, and clients have the option to automatically or manually redefine the data.

The contents of the server are also accessible by any other client onboard who needs the data. The system is run by a system's manager supervising various managers in charge of sessions, plots, tracks, operators etc. All the users can access the data base. A separate data fusion gateway is connected to the data processing component. The man – machine interface is replaced by GUIs (graphic user interfaces). A single console screen is used with overlaid pages to give other information required like e.g. time-bearing plot, contact classification plot, narrow band analysis, ESM inputs, periscope feeds, target motion parameters and so on. With such systems it now becomes easy to present the commanding officer with a fuller tactical picture and enable the submarine track and attack more than one surface target at a time. The vagaries of the sea will, nevertheless, continue to present inaccuracies in data inputs.

Moving to the underwater scene, *a torpedo attack by a submarine on a submarine* is rare but a distinct possibility during hostilities, and is altogether a different kettle of fish. The elements of SSK operations have been described in another chapter and only the torpedo firing aspects are being highlighted here. In a *submarine versus submarine* encounter, owing to both the hunter and the hunted being quieter than surface ships, detection ranges of each other, and reaction times, are relatively small. There is also considerable difficulty experienced in classifying the detected submarine. If one of the two is on an SSK mission, that submarine would have a slight advantage over the other, but only barely so. Even so, taking time for classification is a luxury that is not usually affordable. In fact, in cases where two SSK submarines are looking for each other, detection takes place almost simultaneously at ranges when torpedoes can be immediately fired at each other. The *aim should always be to fire first.* Also, it is prudent to always fire an initial salvo of at least a pair of homing torpedoes at the target.

As in the case of surface targets, the attacking submarine, by virtue of having been deployed for this role, and therefore more alert, must use this advantage to quickly classify the target and work out the target motion parameters. An ideal situation would be one where the other submarine (a conventional submarine) is at periscope, depth charging her batteries. Having classified the target as hostile, the attacker can use one of three methods to launch an assault. An attack after **full preparation** may be carried out by the attacker when he knows he has a distinct detection range advantage over the adversary. If the attacker is not sure of this, he is advised to either carry out an attack after **minimum preparation**, firing with an arbitrary lead angle set on the torpedo in the direction of target movement, or an **urgent attack** at the target's present bearing.

It would be wise to carry out an urgent attack with a pair of homing torpedoes first. The torpedoes are fired at the present bearing of the target. With the release of torpedoes, the element of surprise would have been lost and the target may be expected to take evasive measures and behave in an indiscreet manner. At this stage, active sonar transmissions could be made to get the range of the target and its motion parameters. Thereafter, the next salvo of two homing torpedoes may be fired at the target with updated parameters. The system now used to attack ships is used for SSK attacks also.

Torpedoes are ejected from their torpedo tubes in different ways in different classes of submarines. There is the *air firing* system. Large firing reservoirs (compressed air bottles) and a depth regulated system of firing air release and inboard venting and compensating was perfected during the 1950s, called Dual Pressure Firing Gear, allowed boats to fire from a depth of up to 150 meters without bubbles of air escaping to the surface (and thereby possibly compromising the position of the submarine). In many conventional submarines, this is the system in use. Some electrically propelled torpedoes are not ejected by air and swim out on their own from inside the tubes. The Brit Mk XXI and the USN Mk 37s did this. In nuclear submarines, discharging torpedoes is affected by using the output of *high pressure seawater pumps* right up to the maximum operating depths of the submarine.

OPERATIONAL/TACTICAL MISSILE ATTACKS

"Some ships are designed to sink... others require our assistance."
A submariner

In an earlier chapter we have discussed how the major powers deploy their strategic submarines with a view to launching Weapons of Mass Destruction (WMD) on selected targets in each other's countries, if and when the 'balloon' goes up. They can practically annihilate each other many times over with the stock piles they have even today, after the Cold War. This chapter is confined to missiles fired by submarines in an operational or tactical scenario , and not in a strategic scenario. With modern day weapons becoming more accurate and lethal, there is an ever increasing urge to launch a weapon at the opponent from outside his weapon range, and get away without getting hurt. Tactical and cruise missiles give submarines that edge today.

Many modern-day conventional submarines and SSNs have the ability and option to fire either tactical or cruise missiles or torpedoes from their torpedo tubes. Some of them have up to eight torpedo tubes with reloads in the compartment, which is considered adequate to work out a permutation combination of missiles, dedicated anti-shipping torpedoes, and dedicated anti- submarine torpedoes loaded in the tubes. Some have six torpedo tubes, and have to be selective in the choice of weapons in the tubes and those on the racks. Some have only four torpedo tubes, and the loading configuration then becomes restrictive. A pair of missiles and a pair of 'universal' torpedoes – the types that can be used against ships as well as submarines – become the preferred option in such cases. This restriction is overcome onboard some submarines by having dedicated Vertical Launch Systems (VLS) for cruise missiles separately.

The Exocet, the Harpoon, the Penguin, Novator 3M-54E Klub S cruise missile, the Chinese YJ-8 missile and the likes were designed and manufactured in the era when blue water confrontations had the highest priority over other capabilities. They were anti-ship missiles. Initial target data was fed to the missile located inside a capsule which was located inside the torpedo tube. The capsule containing the missile was then fired from the torpedo tube. When the capsule breached the sea surface, the top of the capsule or the capsule itself was then blown off and the missile launched. High survivability and effectiveness were ensured by their low level, sea-skimming trajectory. The missile switched on her homing head at the appropriate time, acquired the target, and then went for it. The submarine forgot about the missile after launch and carried on about her other business. Constant upgrades have kept these missiles current with no need for replacements.

When sub-sonic missiles are fired at targets at extreme ranges, some of them may require midcourse corrections. Most missiles have intelligent homing heads that are able to differentiate and select the designated target from a group of ships in formation, with a high probability of hit. However, apart from the missile's own accuracy, the success of selectively hitting a target in a group of ships will depend to some extent on the distance between adjacent ships, the dispersion, the point at which homing head of the missile is switched on, and the dimensions of the indicated field and area being searched by the homing head of the missile.

The warheads of missiles have a high destructive capability. The unspent fuels they carry at the time of hit add to the destructive capability as was shown during the Falklands conflict of 1982. The missiles themselves are vulnerable to enemy fire, and every endeavor is made by designers to get the missile to cruise as close to the sea surface as possible to make them impervious to anti-missile strikes. Last stage maneuvers are also incorporated to assist them in getting past anti-missile defenses. With technology and metallurgy catching up with tactical requirements, these sea-skimming sub-sonic tactical missiles are being replaced by missiles with higher speeds, the Indian *'Brahmos'* supersonic missile being one example.

Weapon ranges were generally less than sensor ranges in most weapon firing systems at sea. This made sense as the sensor had first to detect and classify the target before the weapon could be released. However, with the installation of missiles onboard platforms at sea, the opposite happened – weapon ranges became greater than sensor ranges. With it, now came the problem of target identification at ranges beyond the submarine's sensors' ranges of detection. The submarine would naturally want to ensure that the missile hits the selected target and not another platform. Target detection, identification, and classification at sea, beyond the horizon and at ranges beyond own sensor detection ranges, are not easy tasks for a submarine. Of necessity, she would have to rely on an external source or platform to locate, identify, and indicate the target, and the others in the vicinity to the target, so that she can fire her missiles at near maximum ranges. That external platform could be an aircraft, a ship, another submarine, or even a command & control station ashore. It could even be a satellite. Of course, prime requirements for taking the help of an external source would be secure and reliable two-way communications, and accuracy in reporting co-related positions.

The missile firing submarine can use the external platform in one of two ways to bracket her target. She could operate and hold the relaying platform on her radar while the relaying platform, in turn, holds the target as well as the firing submarine on her radar. The relaying platform then relays data based on which the firing submarine inputs target data and launches her missile/s. In the second method, the attacking submarine does not have the relaying platform on her screen and neither does the relaying platform have the attacking submarine on her screen. The relaying platform, however, holds the target on her screen. A common grid system is then used to relay information and hit the target. To maintain the element of 'surprise' when launching a missile attack, one must ensure that the relaying platform is undetectable by the selected target and other units with it. Usually a few units are selected and worked up during peacetime exercises as relaying platforms, so that any of them can be used effectively when required. It is desirable that all missile carrying platforms be trained as relaying platforms. During peacetime exercises, and even during hostilities, care should be taken to ensure that friendly forces are nowhere in the direct or indirect flight path of the missile before launch.

Missiles against moving targets at sea can be fired at the target's present position or the target's future position. If the target is not on a steady course and is altering courses randomly, it is normally recommended that missiles be fired at the target's present position. If the next leg of the target can be predicted with reasonable accuracy, the missile may be fired on its future position. With more sophisticated missiles available with some of the advanced nations wherein accurate target data is continuously fed to the missile, it could be fired at the present or future position. Most missiles have an initial dead range which is the distance they take after launch to settle down. Before firing, and based on the accuracy obtained of the target's position, the time lapse after which the homing head is required to be switched on will be fed to the missile by the Missile Firing Computer (MFC). This is crucial and carefully done so as to deny the adversary time for degrading the performance of the homing head. The homing head's window is opened as late as possible, and for the minimum time. However, if target parameters are not as accurate as one would like them to be, it may be necessary to activate the homing head earlier to cover a larger field and accept possible negative consequences.

To fire a tactical missile on a formation, it is not necessary for the launching submarine to position herself inside the LLAs of the formation. The missile's speed advantage permits firing from anywhere (even the formation's stern sector) as long as the target is within the missile's range. However, it makes tactical sense for a submarine to close in and follow up an attack by cruise missiles with a torpedo attack if the target has not been sunk as a result of the first attack, or if other targets of opportunity are available in the vicinity. If such an intention is there, it would be better for the attacking submarine to position herself within the LLAs of the advancing formation for the missile attack also, bearing in mind that any SSNs in support of the formation may have to be contended with. With the shift in emphasis from 'blue water' to 'from the sea' maritime operations, cruise missiles like the American Tomahawk and the Russian Klub (SS-N-27) gained more importance, and submarine launched versions soon followed the air-launched and surface-launched versions. These cruise missiles are useable in both roles, and have variants to suit different roles.

Objects on land can be targeted more easily as most of them will be static targets whose geographical co-ordinates will be known, and are fixed. These missiles have the ability to fall within meters of the point of aim. A satellite or shore station could also be designated to indicate the target's co-ordinates. **SLCMs** (**S**ubmarine **L**aunched **C**ruise **M**issiles) present the submarine with the ability to hit their targets beyond the horizon at great ranges and thereby increase their tactical strike ranges. Most modern day SLCMs are sub-sonic and have ranges of around 300 nautical miles. They can be fired directly at the target or indirectly by setting way points to pass through, to conceal the submerged firing platform's location. With these kinds of ranges they are able to target large warships, merchantmen, and formations at sea, as also command and control centers, naval harbors, oil dumps, shipyards, dockyards, and the likes, on land. The land attack capability is useful when operating independently or as a part of coalition, expeditionary, forces in low intensity conflicts and littoral warfare. Adequate information is available in the public domain about two well-known types of cruise missiles - the family of Tomahawks and the family of Klub missiles - and is therefore not being repeated in depth here.

All submarines do not enjoy complete freedom when firing or launching their missiles. There are speed and depth restrictions imposed on some of them at the time of launch. For example in her time, the 'Charlie' class Russian submarine could only fire her missiles at speeds between 8 to 12 knots, and at depths not exceeding 130 feet. Some current Chinese submarines have the same restrictions. The deeper you want to fire the missile from, the more protection you need to give to it and the firing system, from the tremendous pressures experienced at those depths. There could be restrictions also imposed due to the torpedo tube system not being allowed to be operated beyond a certain depth. By providing dedicated VLS launchers for cruise missiles, and the high pressure sea water pumps launch system to the torpedo tubes, all restrictions have been overcome in some of the later classes of submarines.

Objects on land can be targeted more easily as most of them will be static targets whose geographical co-ordinates will be known, and are fixed. Alternatively, a satellite or shore station may be designated to indicate the target's co-ordinates.

There are SSGNs like the Russian **'Oscar'** Class nuclear propelled boats that also carry anti-ship cruise missiles with either conventional or nuclear warheads. They can move at sustained high speeds over great distances unlike the conventional boats, and can be tasked with chasing and attacking Carrier Battle Groups at sea. SSNs like the American **'Los Angeles'** class carry cruise missiles that can be launched either through torpedo tubes or through separate dedicated VLS launchers There are four types of these 'Tomahawk' missiles - TASMs (Tomahawk Anti-Ship Missiles), TLAMs-N (Tomahawk Land Attack Missiles-Nuclear), TLAM-C (Tomahawk Land Attack Missiles-Conventional), TLAMs-D (Tomahawk Land Attack Missiles – Conventional, Bomblets). These SSNs have the flexibility to be deployed anywhere and can play an effective part with the cruise missiles they carry.

After the START II treaty was signed, some of the original American SSBNs have been divested of their Weapons of Mass Destruction (WMD) and converted to carry cruise missiles. For example, the U.S. Navy has converted its four oldest Ohio class Trident submarines from an SSBN to SSGN configuration. This was completed between 2002 and 2008. Vertical Launching Systems (VLS) now equip 22 of the 24 missile tubes (which previously held one large nuclear-tipped strategic ballistic missile), with 7 smaller Tomahawk cruise missiles. The 2 remaining tubes have been converted to 'Lock Out Chambers' (LOC) for use by Special Forces. Each of these submarines can now carry 154 Tomahawk missiles. The tubes have also been configured to carry and launch UAVs (Unmanned Airborne Vehicles) or UUVs (Unmanned Underwater Vehicles). The converted Ohio class submarines can now be used as a useful command & control center in the area of operations. The Russian navy, after the collapse of the Soviet Union, fell into disarray and have not been able to make any such purposeful conversions of their SSBNs as yet.

MINE LAYING OPERATIONS

This is a peculiar weapon. You need not even lay it – just declare the field.
The consequential effects on the adversary are almost the same
Anon

Mine warfare is an integral part of maritime warfare that is used to control SLOCs (Sea Lanes Of Communications) and littoral water space, while denying the adversary the use of same. A sea mine is nothing more than an underwater explosive device that is placed in selected water space to wait for and destroy ships or submarines passing through that area. They are normally placed in waters of up to 50 meters in depth against surface ships, and in waters up to 200 meters in depth against submarines. Launched in an inert state, they are 'activated' even before a victim comes their way. Own forces are not required to be around for them to work or be activated. They can be crude in construction and still be very effective. They are much cheaper than other options with the same ends in mind. There are types and types of mines – ground mines, moored mines, contact mines, remotely controlled mines, influence mines, etc. They can be deployed offensively, defensively, or for psychological effect. When used offensively, they can tie down or contain enemy forces in their own waters or harbors, or sink his forces in his shipping routes. When deployed defensively, they can protect own forces and create havens or shelters from a belligerent force. Psychological minefields are usually laid along SLOCs to prevent any support or assistance reaching the adversary from other sources. They are laid thinly over a large area to get random hits over a large water space. The resultant psychological effects are worth the effort.

Sea mines can be laid by aircraft, ships, and submarines. They can also be laid by supposedly innocent vessels like merchantmen, trawlers, and fishing vessels. In fact, just about any platform at sea can lay mines. Submarines have the advantage over all of them of being able to lay mines discreetly in areas far removed from their base or own shores, even under the adversary's nose, in his waters. They can do this independently without the need for covering escorts. They can carry their mines in an active state all the way, or in an inert state to be activated on site before launch. Submarines should therefore be the obvious option for laying offensive or psychological mine fields.

There are limitations on the number of mines a submarine can carry onboard at a time. Two mines can be carried inside each torpedo tube (a few torpedo tubes must, perforce, carry torpedoes for self-defense). Some submarines are limited by the number of torpedo tubes onboard and the stowage space inboard for tube reloads (each rack in the compartment can carry two mines in lieu of one torpedo), while others are restricted by the number of saddles they can carry on the exterior and the number of mines each saddle can carry. A lone submarine can therefore not be expected to establish a dense minefield of significant dimensions. Another disadvantage is that submarines with mine saddles have a slow speed of advance and take time to get to the selected area that is to be mined. The further away it is, the longer they will take to get there.

Once a decision is taken that a submarine is required to lay mines, the areas where minefields could be laid by her should be carefully selected and defined by the submarine planning staff ashore. These areas are normally selected from known and promulgated routes of enemy movements, often at the approaches to their harbors, and in straits and narrows and choke points through which hostile forces have to/are known to pass.

The next step would be to select the type of mines the submarine is required to lay. To make it difficult for the adversary to neutralize the mines, a variety of all types of mines that the submarine can lay may be selected for a given area. With the available intelligence of the area, the way in which the submarine is required to lay her assortment of mines should next be worked out by the staff. They could be laid with the help of pre-determined co-ordinates in pre-determined areas. The staff, having worked out these aspects, should next issue mission orders to the nominated submarine. If intelligence inputs on the area to be mined are adequate, specific orders may be issued on where to lay the mines. If the submarine is required to carry out reconnaissance in the area first, gather intelligence, and then lay mines, the exact location of the mine field in the area may be left to the discretion of the submarine commanding officer to select. In such an event, the effect that the minefield is expected to create needs to be specified to the commanding officer.

If the mission is going to be specified in detail in the orders to the submarine, the area selected should be clearly defined. The time of laying, and the duration permitted to lay the field, should be stated. The recommended ways and means of laying the field, the pattern, and the necessity to watch for any results, if required, should be stated. Information on the disposition of routes used by the adversary as gathered from intelligence, or assumed based on other sources of information; the approaches to them, as also to the adversary's harbors, should be included in the mission orders. All available information and data on the adversary's anti-submarine forces likely to be found in the mine-laying area should form a part of the mission orders. Updates can easily be obtained from satellites and other sources, and these should be signaled to the submarine right up to the moment she reaches the area to be mined. Minefields may be laid as a precursor to other major operations. If that be the case, there would be some degree of urgency to report back successful completion of laying the field and this must be specifically stated in the mission orders.

Armed with mission orders and mines onboard, the submarine then sets out for the designated area while the expectant shore staff waits for developments. On arrival in the area and before laying mines, it would be prudent on the part of the submarine commander to carry out a careful reconnaissance of the area and establish the intensity and pattern of operations of the adversary's ships in the area. If the field is to be laid at the approaches to the adversary's harbor, he should establish the orientation of his 'swept channels' and transit routes of his shipping. He should also ascertain the hydrological, meteorological and navigational conditions in the area. The submarine is now ready with all the information she requires to begin her operations.

Lines of Mines Laid By A Single Submarine

If, for whatever reason, the staff has left the decision of where to lay the mines to the submarine commander, he should now decide the area, the best way to lay the field and its elements, the Starting Point and time of laying each line, group, and single mines. He should decide the speed and depth of the submarine for each run when laying mines. He should also work out the type of maneuvers to be carried out when re-loading mines from racks into the torpedo tubes, taking care to stay clear of the mines just laid and be in position to get back to the correct start point to lay the rest of the mines. Depending on the aim of laying the field, geo-strategic conditions, and conditions in the region, submarines may lay mines uninterrupted, or in staggered rows. Laying is executed secretly, at depth, and at the selected speed. After laying the mine-field, the submarine should leave the area and report completion of task, if required to do so.

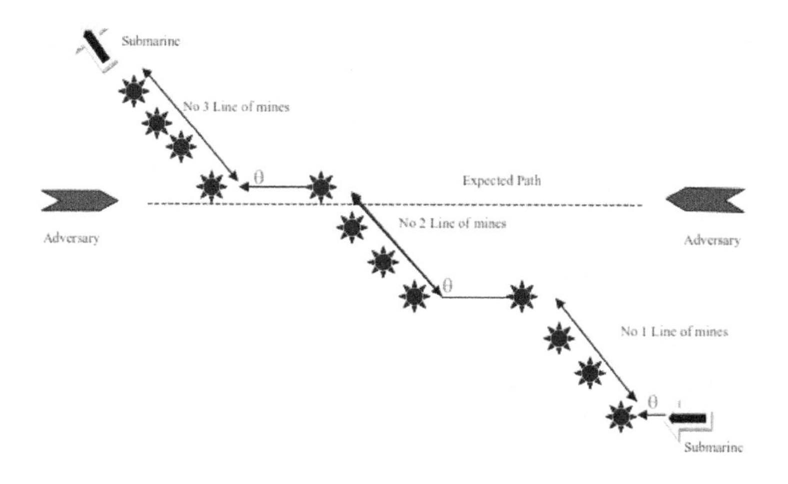

Lines of Mines Laid By A Single Submarine

There are international rules exist like the San Remo Manual of 12 June 1994 and the Haag VIII Convention, urging countries laying mines to follow a moral code of conduct. Those laying mines are also expected to recover or neutralize them after the need or hostilities are over. Recovery or neutralization is a tedious and long drawn out procedure that sometimes takes years and years. Even at the end of that period one is never sure that all of them that were laid have been neutralized. A case in point is the Sounds and Kattegat that link the North Sea with the Baltic Sea. Active mines laid during World War II in these waters were accidentally recovered by fishing vessels and trawlers in their nets even as late as thirty years later. Mines are usually the cheaper weapons of a weaker nation or rogue state fighting a superior force. Massive effort requires to be put in by mine hunters, mine sweepers, ROVs (Remotely Operated Vehicles) etc, to clear a field that has been laid. Some of the nations that have laid these fields are not ever likely to neutralize them after the need has been met.

SUBMARINE COMMUNICATIONS

"Brevity, security, and communicating on need-to-know basis are the keys for submarine communications."

Passing mention has been made in different chapters about communicating from and with a submarine with respect to preserving her attributes of concealment and surprise. This section highlights all aspects of submarine communications, so as to be able to view this subject in the right perspective under different operating conditions.

Post- Cold War, insofar as deployment of nuclear propelled submarines are concerned, there has been a shift in submarine mission emphasis. This shift is from a *sea control* or *sea denial* mission to one of supporting regional conflicts in '*from the sea*' operations, and is becoming increasingly common in sensitive littoral areas in all the oceans. Attack submarines (SSNs and SSGNs) are now participating in littoral operations, joint (SLCM) strikes in co-ordination with surface groups, surveillance to support such operations, intelligence and quick reporting, space and electronic warfare, sea lift/protection and so on. As a consequence, new communications equipment are now in use that provide for complete interoperability with surface groups, timeliness in sending/receiving messages, rapid retargeting of SLCMs, greater reliability, a shared common tactical picture, and robust throughputs. More improved retractable masts are also being progressively introduced on submarines.

Automation and band switching have been brought in to contain the urge to increase manpower to match these increased demands. Data throughput capabilities have been incorporated on most modern submarines to support imagery to allow timely transfer of strike and surveillance missions. New antenna configurations have been designed for nuclear propelled attack submarines operating with expeditionary task forces to envelope all communication requirements, with emphasis on the higher frequency bands and high throughput regimes (SHF, EHF, MDR and UHF). The configuration now provides assured connectivity across the spectrum of conflict. Interoperability with expeditionary task force commanders and the task force, with reliability and high data rates of transfer, is now the primary requirement of modern submarine communications.

SSBNs continue to perform their deterrence role while conventional submarines continue to perform their traditional as well as limited coalition roles mentioned in earlier sections. Both are, however, beneficiaries of some of the new communications packages that have been installed onboard modern attack submarines to meet present day requirements.

Contrary to common belief, conventional submarines communicate with shore authorities, with ships, and amongst themselves, at various times and under various conditions. They communicate via multiple, complementary Radio Frequency (RF) systems, covering nearly all military communication frequencies. There are space restrictions on the type and shape of antennae that a submarine can have onboard. Their material compositions are special to cater for the extreme environmental conditions, and stealth considerations. They also perform differently to antennae onboard ships. Even so, a conventional submarine manages to carry an affordable suite of antennae to provide necessary communications, the IFF responder, EW, and navigation requirements. Reception facilities exist onboard to receive Extreme Low Frequency (ELF) and Very Low Frequency (VLF) signals under water, albeit at a low data rate. Inside the hull, the submarine carries both RF transmitters/receivers and base band suites.

How and when is a submarine called upon to restrict the use of one or the other of her many communication facilities in the interest of stealth? We can, broadly, divide their operations requirements into four general conditions.

The first condition is when a submarine is operating on the surface, or under conditions that could be termed as 'overt' in nature. Showing herself on the surface would offer a high probability of detection. In such an environment, she would be free to communicate in any or all of the channels available at her disposal, and these would normally include EHF MDR (Medium Data Rate - SATCOM), Ultra High Frequency (UHF), Super High Frequency (SHF), Very High Frequency (VHF), High Frequency (HF), VLF and/or ELF frequency bands. She could also operate her radar or simply raise her Electronic Support Measures (ESM) mast for passive reception.

A second condition is when she comes up shallow from deep for the purpose of receiving signals or for any other reason, and is stationary just below the sea surface, in a 'trimmed down' position. In this condition, she would present a low risk of being detected and could raise the necessary antennae to receive EHF MDR (Medium Data Rate - SATCOM), SHF, UHF, VHF, HF, VLF and/or ELF transmissions. Here, again, she could either put up her radar or ESM mast to detect unwelcome presence of others in the area. While in the reception mode she may not be open to danger, transmitting on some these frequencies would present a high risk of compromising her presence and position in both of the conditions just enumerated.

Conventional submarines do come up and operate at periscope depth. This requirement may come up for battery charging or when carrying out covert operations, for example. At periscope depth she would present a low to medium risk of being detected that can be maintained at that level so long as she does not transmit and make herself 'voluble'. Channels of communications open to her with her communication masts up would be in the EHF LDR/HDR(Low/High Data Rate - SATCOM), SHF, UHF, VHF, HF, VLF and/or ELF frequency bands.

Finally there is the continuously submerged mode: the fourth condition that we could term as the *stealth* mode, when she can only receive signals and not make any transmissions. She would not raise any antennae above the sea surface and would, at the most, trail a receiver antenna behind her that would come shallow but not be visible on the surface, and would be able to receive VLF or ELF messages at a low data rate at depth. Communications with ships or other submarines with the help of sonar could offer her another alternative.

Against this back drop, the old requirements, for conventional submarines in particular, have not ceased, and the systems in place continue to be available when boats proceed on independent missions. Communications with conventional submarines from ashore are designed and formulated in such a manner as to ensure timely receipt of information required by the boat at sea in order for her to perform her assigned mission in the best possible manner. Tactical considerations or the presence of 'strangers' in the vicinity may require that the submarine remain deep for some hours to come, from where she would not be in a position to receive signals. A particular diesel-electric submarine commander may want to remain deep throughout the day, and come shallow to receive signals only during dark hours. Another may find it uncomfortable to stream the antenna in the area of operations and choose to receive signals only when he comes up to periscope depth, the timing of which cannot be predicted. The towed wire antenna may be non-operational. Then what? Signals must still get across to the submarine, and that too before they become irrelevant. The organization and system in place ensure that no matter what the regime of operation of the submarine at sea, the required messages get across in time for her to react. Elaborate and exclusive arrangements are therefore established for submarine communications, with security of information and brevity holding the key.

OPERATING IN COMPLEX ENVIRONMENTS

"It is important to know who is who, who is to be hit, and who is not to be hit."

- Anon

It is an uncomplicated situation when navies are able to fight against each other without interference from, or by, a third party. Anyone in those waters that are not a part of your own navy can be taken to be the adversary and dealt with accordingly. No merchantmen, no third navy around, to complicate the picture. In recent times, the Falklands War may be considered to be one war that was fought under such conditions. It was fought far removed from shipping traffic or maritime interests of other nations. By shutting the Suez Canal down in the nineteen sixties, the Arab-Israel Wars were fought in that region with no interference from 'outsiders' to complicate matters too. However, most of the other naval engagements that have taken place (including the Vietnam War) since World War II, have been fraught with the problem of 'others', not involved in the business of war, transiting or making way through the battle zone and expecting both warring sides not to harm them. And, why not? - Aren't the waters they are transiting through 'international waters' that can be used by anyone? Some of them came to grief with this logic.

As stated in an earlier section, a submarine's transit route may be planned through waters where ships of all types, belonging to more than one nation, ply. As long as 'evasion' is the policy when plying through these waters, no problems would be posed. However, the same waters may be used by the adversary too, and herein lie the problems. In international waters, along shipping routes that permit innocent passage by law, how does one differentiate the neutral from the opposition and actively take on an adversary, especially when there are a number of neutrals milling around?

Consider a hypothetical, but not inconceivable, situation of two neighboring nations at war, with forces deployed against each other. As part of their offensive operations, both of them have deployed two submarines each. One of them shares a border and coastline with a third country that is facing threatening postures against it by coalition forces. In other words, there are two hostile scenarios that are not inter-related, but they share common waters as part of their area of operations. The coalition forces have deployed an SSN in support of land attack operations, and the opposing nation has deployed SSK submarines against them. We now have a situation where four to five submarines are possibly milling around in the same area, not to mention the number of warships going about their business in the same waters.

Such complex situations could be confronted in the **North Arabian Sea**, for example, from where practically the whole world buys its oil. Tankers of different nations, as well as warships protecting their tankers, are plying through such waters practically twenty four hours a day. The English Channel, the Straits of Gibraltar, the Malacca Strait, and South China Sea are all examples of areas through which ships of the whole world ply. To have a hostile situation in such waters between two warring nations can complicate their lives and also the lives of others using those waters. Submarines find themselves in the middle of such environments very often. Stretching it a bit further, submarines of third nations sent on intelligence gathering or reconnaissance missions in the neighborhood have to contend with two warring nations in the same waters, having a go at each other! One can think of many such complex situations in the troubled world of today.

Two important issues arise from these situations – two situations that will confront submarine commanding officers operating their submarines in such waters. The first one is the requirement for *proper identification*. When operating in neutral, international, waters it is important to 'separate the wheat from the chaff'. To do this successfully in a dived state, without coming up shallow for identification, it is necessary to have a good data bank with as many details as possible stored in it that will assist the submarine in identifying who is who. As a first step, the data bank must have details of own ships – warships, submarines, coast guard ships, research vessels, ships plying between off-shore installations and ports, merchant marine coastal ships, merchant marine main fleet, et al. Compilation of this kind of data bank is a perennial peace-time, round the clock activity that cannot wait for impending hostilities to show up, to work on. Those in the business of doing this know just how complex gathering such data is. With such information in the data bank, it would be possible to identify own ships and submarines as and when the submarine comes across them.

Another peace time activity is to gather similar information on all craft belonging to likely adversaries and friendly nations in your area of operations. This is even more difficult than gathering data of your own forces. All submarines, when out on sea sorties, must endeavor to gain such information, store it, and update it as often as opportunity presents itself. What is left is data of all other ships of other nations plying in the submarine's areas of interest. By a process of elimination with the first two categories of data in hand, it would be possible to identify these neutrals and leave them alone. Needless to say that, the system for acquiring and stowing such data must be the best that can be got and installed onboard.

The second requirement is for the submarine to sail out into such complex environments with proper *Rules Of Engagement* (ROE) both during peace and during hostilities. ROE are normally issued to ships and submarines proceeding to sea by higher naval authorities ashore with their government's stamp of approval. They are supposed to specify in detail the circumstances under which hostile acts may be permitted at sea. The ROE are supposed to be *comprehensive* instructions enveloping a wide range of contingencies that a commanding officer may be generally faced with at sea. However, the ROE may also be framed for a *particular operation*, in which case it *can be more specific*. A specific ROE endows the commanding officer of the submarine with greater confidence and enables him to deal with threats boldly and more effectively.

There is the well-publicized incident of the *USS Stark* that took place in the Persian Gulf on 18 March 1987. A Mirage F-1 fighter of the Iraqi Air Force fired two Exocet AM39 missiles that hit and holed her. The ROE with the Commanding Officer was vague enough for him to allow the aircraft to come within weapon release range, and the ship suffered the consequences. Lack of proper identification resulted in the *USS Vincennes* bringing down an Iranian Commercial aircraft and killing all onboard, on 03 July 1988 – again in the Persian Gulf. Proper identification and a more specific set of ROEs may well have avoided both incidents and resulted in the Mirage F-1 being shot down and the Iranian commercial airliner getting away.

Are there any other ways of going around such problems? There are many alternative options worth considering by planners ashore before deploying ships and submarines during periods of hostility. Some of these could be:-

- Declaring minefields in specific areas to keep ships and submarines away from there. They may be offensive or defensive in nature. The minefields may or may not be laid, but the announcement must carry with it a certain amount of conviction! Who is to know or chance an entry?
- Declaring a war zone and warning all neutrals to keep away from the zone.
- Declaring a blockade of a certain port or facility and advising neutrals to stay away.
- Shutting down a passageway (like the Suez Canal) thereby forcing neutrals to take an alternate route.

When joint expeditionary forces operate against a common adversary, it is desirable that a common ROE be promulgated and followed. From the point of view of commanders at sea, it is also desirable that these ROEs be as specific as possible. It makes their life that much easier. Today, individual navies act in accordance with ROEs promulgated by their respective governments. This can cause complications in joint operations. In pursuance of trying to have a common ROE for joint expeditionary forces, governments will tend to promulgate vague ROEs that can cover as many situations likely to be confronted at sea as possible, much to the dissatisfaction of commanding officers at sea. These ROEs will still, however, have minor differences. With operations concerning coalition forces on the rise, there is a need to find common solutions to complicated situations beforehand, and definitely before issuing ROEs.

SUBMARINE RESCUE

"Submariners are a special brotherhood, either all come to the surface or no one does."

- Anon

All submariners train to survive and escape from a sunken submarine or to be rescued with external help. All submariners! Without qualifying in this art of escaping from a sunken submarine, no submariner should be embarked to put out to sea in the boat. To know that one can be rescued if any untoward incident or accident compelled one to abandon the submarine is a comforting thought. It does much to the morale of the submariner. Some navies have rescue facilities, while others depend on their allies to provide such facilities, should the need arise. Still others, who have no rescue facilities or allies to help, hope to receive help on humanitarian grounds from anybody and everybody if and when the moment of truth confronts them. On humanitarian grounds, others have helped and will continue to do so, the attempts on the Kursk being one such example.

Submarine rescue, though practiced, is the least discussed topic among submariners. The only time it is discussed is during the escape courses or refresher courses that are conducted on a regular basis. As it is, the profession brings with it its own risks. Those risks, and a catastrophe like a sunken submarine, are thoughts that submariners least want to dwell upon. Yet they train thoroughly for such an eventuality. Do they believe in this business of rescue? Some do and some don't! Those skeptics, who don't, keep their opinions to themselves and let others believe in it. Certainly, incidents of submarine disasters with 'no survivors' that publicly appear every once in a while, do not help towards believing in this art.

The *Kursk* was a state-of-the-art Russian submarine and one of the best in its class at the time when disaster struck it. It was about 300 feet long, and after the disaster, settled in a little under 500 feet of water. Stood end on, the stern or bow would have been less than a mere 200 feet below the sea surface. All the external help that was made available could not result in pulling a single man out alive. Neither did the internal equipment for the crew and their competency in escaping by themselves that is practiced so often. The

PNS Ghazi sank in shallow waters off the fairway buoy leading to the entrance to the Indian port of Visakhapatnam, after an explosion in 1971. There were no survivors. One may argue that a war was on at that time, and external help would not have been forthcoming. But none of the crew members managed to escape by themselves as they are trained to do. The waters were shallow enough for them to attempt it. The *USS Thresher* and the *USS Scorpion* both sank in very deep waters where even locating them took months. The Indian Navy's INS Sindhurakshak imploded alongside the pier and sank just a few meters with its fin top still visible. There were no survivors. There couldn't have been. The list goes on ....

If one looks at the flip side of the coin, there have been many incidents where crew members have been rescued from stricken submarines at sea. However, in most cases, these have been after the submarines have surfaced. Most of the boats eventually sank.

The chances of survival begin from the very beginning – from the design and construction stages themselves. A good reserve *buoyancy*, checks and balances in the design against flooding due to external explosions and related hydraulic buffeting, doublers on hull fittings, exacting specifications for as many minor fitments onboard as possible, installation of flame resistant materials to the maximum extent possible, a good fire suppressant system onboard, high water tight integrity in each compartment, powerful pumps for pumping out water from the flooding compartment at a fast rate, alternate sources of power for controlling the submarine's control surfaces, propeller/propulsor, and equipment considered vital for her survival, a system for removing/absorbing poisonous gases and restoring breathable air, etc. The list is endless.

What is being highlighted here is that apart from all considerations that go into a submarine design to help it to fight and survive in a high pressure natural environment that the undersea water space offers, a reserve is required to help it to survive accidents. Designs concentrate on survivability after an external hit during war. An internal explosion does not get the attention it deserves. Historically, there have been more disasters due to *fire* than flooding inside the pressure hull of submarines. Electrical fires can lead to blinding flashes and blasts that stun and cause fatalities in their immediate vicinity in confined spaces. Poisonous gases in the compartment render the crew in the vicinity unconscious and useless.

When sea water comes into contact with lead acid batteries that exist in large numbers inside conventional submarines, chlorine, stibene, and other poisonous gases escape into the air, causing extreme discomfort to the crew, often leaving them inactive. These have to be addressed at the design stage.

The next chances of survival rest on correct drills by crew members to counter fire and flooding. These have to be exercised as many times and as often as opportunities present themselves. These drills should form a part of work up that a crew has to go through even before the submarine puts out to sea, and also during work up at sea. It is recommended that fully worked up crews continue with exercising these drills as often as possible, and on a continuous basis. Some imagination is required when simulating incidents of fire and flooding onboard, with a different situation battled against each time. To remove subjectivity, and to infuse new incidents, it is worth considering getting an external organization to come onboard and put the crew through damage control drill paces, or work out intensively in Damage Control Simulators if available ashore for this purpose.

The *USS San Francisco*, cruising at flank speed at a depth of 525 feet in the Pacific Ocean, rammed into a submerged hillock not shown on its chart, and managed to get home under own power for repairs. This incident got coverage in the media and on the internet. There was just one loss of life, and only 24 were injured. This speaks volumes for the boat's design. The collision (or grounding, if you will) could have been fatal.

When a submarine is designed, arrangements for the crew to escape from it by themselves (if such an eventuality ever comes up) are catered for. The crew can *escape through the torpedo tubes*. It is a simple procedure. They would be required to wear special escape suits with breathing apparatus, and crawl into each empty torpedo tube, two at a time. Getting into a 533 mm diameter tube, two at a time, is the ultimate test for claustrophobia!

The first in has to crawl his way right up to the Bow Cap of the tube. The second one follows, and once both are in, the Rear Door is shut and the tube flooded with sea water. The pressure in the tube and outside the bow cap (open sea) is then equalized. The bow cap is then opened and the two of them swim out. The bow cap is shut and the tube drained of water thereafter, in preparation for the next two to crawl in. The ones who have crawled out have to follow a drill of 'stops' and 'moves' at various depths, all the way up to the surface depending on the depth from which they are escaping. For this they have a rope tied to the Bow Cap (by the first man who comes out) which they must hold on to while moving up. This is to avoid decompression sickness. If they are fortunate to have a ship waiting on top with a decompression chamber onboard, they can do a free ascent all the way to the sea surface, and quickly get into the decompression chamber where they will be put under pressure and slowly decompressed till they are free to come out. The whole procedure is traumatic. The alternate option of remaining onboard the submarine and slowly dying is certainly worse, and it is this thought that spurs them on to escape.

Then there is the other option of *swimming out through the hatches leading out of the pressure hull - forward torpedo loading hatch, the aft hatch, and/or the sail hatch.* There are set procedures to accomplish these tasks and submariners know them. The point is, not all these may be operable or available to escape. What is operable and what is not operable would depend on the type of damage the submarine has suffered, which compartments are flooded and which are negotiable, and the depth and attitude at which she is lying on the bottom of the ocean (given that the bottom is not always flat). The explosion in the bows killed all personnel in that compartment onboard the *Kursk* instantaneously. It instantaneously killed all in the torpedo compartment onboard the *PNS Ghazi* and *INS Sindhurakshak* too. The compartments were blown open and fully flooded. It was not available for escape to the others onboard, who opted to move aft to survive, in the hope that external aid would come before they ultimately ran out of air to breathe.

The crew can escape by themselves and without external aid from depths of around 120 meters or less. Of course, they would all have to wear breathing apparatus and escape suits. If the submarine is lying at a greater depth, then a specialized submarine rescue vessel with associated equipment or a DSRV (Deep Sea Rescue Vessel) would be required to get the crew out of the boat. The specialized submarine rescue vessel may be equipped with a rescue bell that can be lowered over one of the submarine escape hatches. Once the bell sits on the hatch coamings, the hatch can be opened and the men transferred into the bell, to be transported to the surface. The DSRV similarly sits on the hatch coamings to transfer personnel into it and then transports them up to the surface. This form of rescue would be possible if the angle and attitude at which the submarine is lying permits the bell or the DSRV to sit on the hatch coamings and lock on, rendering the space between them and the escape hatch watertight. If this is not possible, then the 'dry' escape would have to be abandoned and a 'wet' escape into the bell or the DSRV attempted. There are submarines like the HDW designed Indian *'Shishumar'* class that have their own *'rescue sphere'* as an integral part of the submarine. Some Russian submarines have them too. The *Rescue Sphere* is released from the submarine, from depth, with the entire crew in it, without any external aid. Once it pops up on the surface of the sea the crew can be rescued.

There are so many 'ifs' and 'buts' involved in submarine rescue. Each situation will be different to the previous one, with all of them equally challenging. Time is of essence in all rescue efforts as breathable air onboard the submarine is limited, and running out of oxygen can asphyxiate the crew. Any damage to the nuclear plant can spread contamination in the boat and that is another matter of serious concern that has to be handled differently. There is a large element of luck involved in executing a successful rescue from a sunken submarine. Luck! - Together with proper drills, and help from the Almighty.

REMOTELY OPERATED VEHICLES FROM SUBMARINES

"To be perfectly secret, one must be so by nature, not by obligation"
Montaigne

During the last fifty years or so, two important concepts have come into being, aided by the progress that science has made, to make them possible. For one, human life has come to be valued more greatly. Therefore more and more thought is constantly being put in to minimize loss of lives during battlefield encounters. Secondly, technology has made it possible for detection systems and weapons to reach out to greater ranges and thereby achieve a 'kill' from over-the-horizon ranges and from beyond detection ranges of the opponent or target with pin-point accuracy. Military satellites are constantly in use during peace for intelligence gathering and surveillance, and equally in use during periods of hostilities to direct ordnance on to targets. Through all this is born the concept of using unmanned vehicles to detect, report, and strike targets with punitive effect. Today, it is possible to sit in an armchair with impunity and control drones from far away United States to strike targets in Afghanistan with these very platforms, with telling results. The expensive aircraft and the vulnerable pilot are kept away from risk and harm and used minimally. This concept of using Remotely Operated Vehicles (ROVs) or Autonomously Operated Vehicles (AOVs) in areas of conflict is very much a part of tactics employed by modern armed forces. At sea, both the airborne version and the undersea versions are deployed, with the twin-fold aim of accomplishing tasks from beyond the enemy's range, preserving own lives onboard, and all this at lesser cost.

This section is going to deal with the latest attempts that are being made to make submarines more 'unreachable' to the opposing force. They are being equipped with *Unmanned Undersea Vehicles* (UUVs), also referred to as underwater drones, which are designed to move away from the parent platform and perform roles, at some distance, on behalf or at the behest of the submarine. UUVs may be defined as submersibles that are self-propelled and operate autonomously (pre-programmed or real-time adaptive mission control) or under minimal supervisory remote control without any personnel onboard. They are platforms that can be launched from ships or submarines. When they are programmed to operate by themselves like robots, they are referred to as *Autonomous Underwater Vehicles* (AUVs). When they are operated remotely by an operator either from the launch platform or any other platform, through tethered data links or fiber optic wires, they are termed as Remotely Operated Underwater Vehicles (ROVs). The use of such platforms gives the submarine greater opportunities for deception, secrecy of operations, access hitherto unapproachable waters, and immunity from damage. Technologically, it is conceivable that AUVs be launched to pick up, identify, track, and destroy a target, and return to the launch platform as part of large scale operations. Development needs to follow.

Deploying UUVs from submarines has the following advantages:-

- Risk reduction to the submarine and its personnel.
- Conduct of operations with low profile, and less susceptibility of being detected.
- They could be launched by one platform, deployed, and recovered by any other platform
    - having similar facilities.
- Can be deployed and operated in bad weather conditions that do not play a part
    - under water and thereby affect the UUV.
- They are cost effectively cheaper to deploy in specific areas than the parent platform.

UUVs could be designed, programmed, and deployed to perform any of the following roles:-

- Surveillance, reconnaissance, and intelligence gathering.
- Monitoring traffic off enemy ports and disseminating information to own forces.
- Mine Detection, mine destruction, and mine laying operations.
- As an effective tool in Anti-submarine Warfare Operations. Track submarines during peace time.
- Gathering hydrographic data in areas inaccessible to submarines, and close to the adversary's harbors.

- Delivering ordnance and destroying targets.
- As roving 'watch-dogs' in defensive roles in protection of off-shore assets and home ports.
- As communication and computer jammers.
- Operational deployment in shallow waters not accessible to submarines.
- Sea-floor mapping.
- A force multiplier.
- Any other roles where, by deploying UUVs and achieving the task, the safety of the submarine is enhanced.

AUVs and ROVs come in different sizes and shapes, and with differing mission specific roles, and may be broadly classified by size and weight into four categories:-
- The portable variety that can easily be handled by personnel, both inside the launch platform and outside it. Many of these already exist, and are commercially available.
- The light weight variety that can be carried by midget submarines (and light surface craft).
- The heavy weight variety that can be carried by SSNs and conventional ocean-going submarines. They are generally of the same diameter as torpedo tubes where they are stored, and from where they are launched.
- The large variety that can be carried by SSBNs and SSGNs in external pods.

They can also be categorized based on their mission capabilities.

The United States Navy has taken a lead in the development and use of UUVs. In the year 2004 they unveiled an unclassified version of "US Navy's Sea Power 21 UUV Master Plan" which declared their intent on which way they were going.

In April 2015, they deployed their first underwater drone from a Virginia Class SSN at sea. It was the REMUS 600 Unmanned Under water Vehicle launched from an 11-meter long detachable/removable module on the Virginia-class submarines called the dry deck shelter which can launch divers and UUVs while submerged. The REMUS 600 is a 500-pound, 3.25-meter long UUV equipped with dual-frequency side-scanning sonar technology, synthetic aperture sonar,

A UUV BEING DETACHED FROM THE PARENT SUBMARINE

acoustic imaging, video cameras and GPS devices that has 20-hour endurance and can operate up to a depth of 600 meters. It is similar to the BLUEFIN 21 that was used to scan the ocean floor for the wreckage of the downed Malaysian airliner in 2014. The US Navy plans to use Commercial Off-The-Shelf (COTS) technology UUVs at the first instance, and modify them to suit military roles. The Russian and Chinese navies cannot be far behind in this field.

UUV technology is not maturing as quickly as UAVs. This is because of the problem of communicating with UUVs. Radio signals cannot penetrate the oceans and acoustic signals take too long. Even while this problem is being battled with, and it is not insurmountable, the first many UUVs will almost certainly have to be completely autonomous. Once this problem is overcome, UUVs are going to be big game changers in undersea warfare. One expert in the field has even predicted that with sophisticated UUVs as part of the inventory of submarines, future boats will have to be bigger and operate like submerged aircraft carriers, further out at sea. That day may not be far away.

CRYSTAL-BALL GAZING (THE FUTURE)

"These dolphins, once you pin them on your chest, leave deep marks, right over your heart, long after the uniforms have been put away."

So what does the future hold for submarine operations in terms of technology and design, dictated by need? Today's world is a *uni - polar* world. It is also a shrinking world. Global access in every field is becoming easier and easier. Contrary to common belief, the end of the Cold War has made the world a more dangerous place to live in. That is the truth of the matter. The only superpower, with no foreseeable threat to it in the near future, is looking at her maritime role in an entirely new light. She, and all other large navies, have downsized after the Cold War and gone in for improved quality in lieu of quantity. Unable to police the entire waters of the world by herself, she is trying to identify suitable nations and navies all around the globe to assist her towards this end. At the moment, the preferred option is for joint expeditionary forces to assist her in carrying out her self-assigned worldwide missions. As a first step, she is carrying out bi-lateral exercises with navies of other nations in all the oceans in an effort to understand their concepts of operations and their thought processes. Other navies are also engaging in joint, bi-lateral, exercises amongst themselves to find a way for inter-operability. The lone superpower has shifted her emphasis from a 'blue water' navy to a 'from the sea' navy, with littoral warfare taking the higher priority, although latest indications are that she has plans to enhance her 'blue water' capability again.

In this new scheme of things, submarines are expected to play an important role. They are being modified to suit new roles being assigned to them. As submarines are proving to be costlier to build with time, to gain a little in one direction, some compromises in another direction are being made to keep costs manageable. Some of the older boats have been modified to take on present role requirements by the lone super power, and as long as she is not seriously challenged, this trend will continue. In her submarine 'new build' programs the superpower is seeking multi-role capabilities, better communications suites, and improvements in speeds with some compromise on acoustics requirements. Hulls are being modified to embark additional equipment outside the pressure hull. They are also being modified inside so as to be able to launch land-attack missiles. Some of the 'near superpower status' nations are also following suit. Affordable and available technology is making this possible.

Let us for a moment look at the efforts and plans being put in place by these navies for submarines as a platform for the future. While maintaining a minimum force level of SSBNs for strategic balance, the United States Navy is building new SSNs for battle space dominance as well as for open ocean missions, and also converting some of her SSBNs to SSGNs for these tasks. The United States Navy is bringing down her number of SSBNs to 14. She has converted 4 of her 'Ohio' Class SSBNs to SSGNs to carry Tomahawk Land Attack Cruise Missiles (TLAM), and Navy Seals with their special craft and/or a mini sub piggy back for Special Forces. She will also be converting 2 *'Seawolf'* class, 23 improved *'Los Angeles'* class, and 8 modified *'Los Angeles'* class submarines to SSGNs that will have a multi-purpose role. She plans to maintain a level of 21 SSNs (20 *'Los Angeles'*, and 1 *'Sturgeon'* class) and expand their roles to take on responsibilities in littoral warfare and thereby increase their contributions to joint expeditionary forces. The new *'Virginia'* class SSNs will progressively replace the *'Los Angeles'* class submarines. Present plans are to build 18 of them. They are being designed for:-

- covert strikes by launching TLAMs from vertical launchers and Torpedo Tubes,
- Anti-Submarine Warfare (ASW) with an advanced combat system and flexible payload of torpedoes,
- Carrier Battle Group Support (CBGS) with advanced electronic sensors and communication equipment,
- covert intelligence operations,
- covert mine laying operations,
- special operations,
- surveillance and reconnaissance.

The Russian navy went through a horrific low after the breakup of the Soviet Union and is slowly on her way to recovery. She has a force level of 14 SSBNs operational – 6 *'Delta IV'* class, 6 *'Delta III'* class, and 2 *'Typhoon'* class submarines. Responding to the downsizing efforts by the US Navy, indications are that she plans to maintain a force level of 12 SSBNs in the foreseeable future. She also has 19 SSNs (9 *'Akula'* class, 5 *'Victor III'* class, 1 *'Sierra'* class, 1 *'Yankee Notch'* class, 1 *'Yankee'* class, and 3 *'Uniform'* class submarines) and 6 SSGNs (*'Oscar II'* class submarines). It is being reported that she intends producing 20 multi-purpose *'Akula II'* class submarines and the new, third generation, *'Severodvinsk'* or *'Yasen'* class submarines (four operational by 2016), to replace the older SSNs as and when they get past their prime.

The French Navy operates 4 SSBNs (2 *'Triomphant'* class and 2 *'L'Inflexible'* Class submarines) and 6 SSNs of the *'Rubis'* Class. Her declared future build program is for 4 SSBNs of the *'Triomphant'* class.

The Royal Navy currently operates 4 SSBNs of the *'Vanguard'* class and 12 SSNs (5 'Swiftsure' class and 7 *'Trafalgar'* class submarines). She plans to convert some *'Trafalgar'* class submarines to carry TLAMs for multi-role tasking, and to construct 3 *'Astute'* class submarines plus another 3 SSNs at a later stage. The SSNs will carry TLAMs.

The Chinese Navy [PLA (N)] operates 1 SSBN of the *'Xia'* class – a derivative of the *'Han'* class. She also operates 5 SSNs of the 'Han' class and 1 SSGN which is a modified *'Romeo'* class diesel – electric submarine. Apart from these, she has acquired and operates *'Kilo'* class submarines ex – Russia, as also a host of locally built *'Romeo'* class submarines that are progressively being replaced by their modern, indigenous *'Yuan'* Class diesel electric boats with AIP. The Chinese are not very forthcoming with their future plans, but as far as is known, they plan to build another SSBN and 6 to 8 SSNs that will be replacements for the 'Han' class (similar to Russian second generation submarines). These new SSNs will be equipped with torpedoes, ASW missiles, anti-shipping missiles, and land attack cruise missiles. The Chinese Navy is showing every intention of moving their submarines away from their shores, into the Pacific and Indian Oceans for operational missions.

As has already been stated, ASEAN nations are operating around 20 submarines in the South China Sea to counter Chinese claims in the region. These are all recent acquisitions.

CHINESE XIA CLASS SSBN

Mention must be made of Israel's conventional submarines separately here, as they are also a force to reckon with in the Arab-Israel world of West Asia. Their 3 *'Dolphin'* class conventional submarines of German design are believed to be armed with nuclear armed (200 Kgs nuclear head) cruise missiles. They also have 3 *'Gal'* class conventional submarines of UK design. Their future plans for submarines are not known, but they intend to retain their nuclear second strike capability from the sea at all costs.

There are many other navies around the world operating conventional submarines (details can be found in Jane's Fighting Ships). They are cheaper options than ships to acquire, and so interest in their purchase or construction will continue with multi-purpose rather than specialized capabilities. This may seem to contradict what has been stated in an earlier chapter that submarines are built with specialized requirements in mind. True! An all-rounder can never match a specialist and there will be some compromises that will have to be made when seeking an all-round capability.

Interoperability with ships and aircraft of joint expeditionary forces is being increasingly demanded. That seems to be the trend, and the direction in which submarines are going. There is urgency on the part of some nations to acquire a missile firing capability, while others having this capability are looking for more lethal warheads on these missiles. While most of these submarines are there to look after the interests of the parent nation, they are also looking forward to making important contributions as a part of joint expeditionary forces, if and when called upon to do so.

Technology will dictate changes in design and performance capabilities of the submarine. If one or the other lags, the effectiveness of the platform to meet the requirements of that era may be seriously curtailed. There is the urge to increase detection ranges; the urge to go deeper to evade detection and anti-submarine weapons; the urge to move faster but quieter; the urge to reduce manpower; the urge to use UUVs; and the urge to cut costs. Technology continues to make gains in the first five directions, but is held in check by the last requirement. Compromises result. The invention of a deep anti-submarine weapon will propel staff requirements to stress on a deeper diving capability at the cost of compromises elsewhere. To avoid being easily detected more efforts to make the boat quieter or 'invisible' will have to be put in. To get increased detection ranges, larger sonar equipment will have to go onboard, which may increase size and corresponding requirements. To get higher speeds a thinner/lighter hull and corresponding compromises in operating depths may have to be accepted. The progress made in an anti-submarine field will always be a cause for opting for a better counter to it in the submarine field. The future will continue to be a cat and mouse game as it always has been.

Deployment strategies and submarine roles should and will naturally evolve from the requirements of that period in time, and will depend heavily on the technology and the submarine support organization available. As already stated, emphasis on requirements has shifted from a 'blue water' requirement to a 'from the sea' requirement. As everything in this world follows a general cyclic pattern, the future may well see a shift in emphasis back to a 'blue water' requirement with corresponding changes in submarine design, capabilities, and the business of undersea warfare.

All that has been stated relates to submarines operating in the operational and tactical levels. What of the Strategic Level? In the foreseeable future, so long as a nuclear war threat persists, SSBNs of nations will continue to carry WMDs aimed at each other's' cities and vital assets. Their deployment areas are going to be nearer and nearer their own ports with increased ranges and accuracy available to their missiles. Deployment in 'bastions' at home may prove more economical in the long run. The operating costs will come down and increased efforts to tail them with SSNs will make the adversary's efforts more strenuous and costly. More and more nations will acquire this capability.

Will unmanned underwater vehicles (UUVs) replace manned submarines in the years ahead? There are increasing efforts being put in to replace man with robots for specific tasks, where the risks to human life are heavy. These efforts will continue and they will definitely make certain tasks easier to be executed. However, the on-the-spot assessments and decision-making levels that are required of a submarine commander at sea preclude his ever being replaced by a robot, even if the inputs for decision-making are made available to the submarine operating staff ashore who can convey orders and decisions to the robot at sea. The time lag in communications will just not be acceptable.

Submarines, as an effective weapon of war, will continue to be a prime requirement of any navy looking for sharp teeth to be effective. Their deployments through peace and war will be a continued effort. They will hold the advantage in maritime affairs so long as they remain discreet. Any compromise on their assets will put them at a distinct disadvantage, till they seize the initiative again.

Finally, there is this question about the men who man these boats. So long as volunteers and manpower continue to be available to man these boats, they will be a potent force. When navies are faced with manpower shortages due to one reason or the other, the submarine fleet and the navy will suffer. A case in point is the Australian Navy in the first decade of this century. An exodus from the submarine cadre grounded a good part of the submarine fleet that should, ordinarily, have been sailing. It has also affected their build program. Volunteers of the right quality who are properly trained and are highly motivated are required to man these boats. They remain, will remain, the most important assets of undersea warfare now and forever.

EPILOGUE

"Without courage, you might as well not be in it. You've got to have courage--moral courage, physical courage--and honor. Honor means telling the truth even when it might not be to your advantage"
Capt. Charles W Rush Jr.

There is a certain fascination, allure, or intrigue if you will, about how warfare involving two unseen opposing units is conducted. From a scenario where face-to- face combat was the order of the day, the scene has now shifted to relying on instrumentation and external inputs to press a button, to release a weapon, and to ensure the extermination of the adversary without even knowing what he looks like. This now happens in the air, on land, and at sea, and is a common way of waging war with the kind of technology and platforms in the inventories of nations today. 'Fascination', 'allures', or 'intrigue' for historians, those delving in military history, and for those who are not directly involved as the attacker or the target! For the actual participants, the feelings could be quite different and varied. Those entering the arena may do so with complex emotions of excitement, anxiety, and fear. Excitement! - Because the opportunity to put into practice, all that was taught, is about to come one's way. 'Anxiety' manifests as a nagging worry – where is the enemy? What is he going to do? When is he going to attack? When are we going to attack? Who will get the first opportunity? - And so on. Fear! - It is that feeling that is commonly associated with pre-battle tensions and butterflies in the stomach. While the 'subconscious' helps to carry on mechanically with drills and procedures practiced again and again so often in classrooms, simulators, in the air, at sea, or on land in peacetime, the 'conscious' reminds one that this is the real thing and at any moment you can be hit, and that too from any direction by an unseen adversary. Weapons of today are capable of taking a detour before hitting the target in order to conceal the actual direction from which they are released. Ballistic weapons shoot up into space and re-enter the atmosphere to strike from above with warnings of little consequence. The lethality of modern weapons covers large areas with decisive finality. In strategic scenarios, where weapons are delivered onto their targets from very great distances, the attacker is excited while the target is oblivious of when he is going to be hit, and therefore without any emotions till he is informed or comes to know that the weapon is on its way. By then it may be too late anyway to react positively with any desired effect!

It is often said that in the fog of war, the side that makes the lesser mistakes comes out on top. Hence the need to be better than the opposition, in every which way you look at the business of war and that includes the domain of undersea warfare. The man behind the gun, translated to 'the man behind the periscope' makes a big difference in the war in inner space. That man – the commanding officer – carries not only the onerous task of successfully completing the mission assigned to him, but also of ensuring the safety of the many men who form a part of his crew. That man is not born overnight to be all that he is supposed to be. He emerges after years of living and working in the environment that submarines provide, working his way up from the lowest ranked officer to the Captain. He is 'molded', in a way, to fit that role. He goes through some rigorous training at various stages to achieve the standards he is required to, and then has to further undergo a course – what is termed by the Royal Navy as the 'Perisher', or its equivalent – and successfully complete it before he is entrusted with the responsibility of leading his men underwater in harm's way. Emotionally, psychologically, and professionally, he has to be the best man available for the job.

Human emotions, the psychological aspects, the nature and character of opposing leaders conducting the operations, the quality of training, and level of competence are all decisive unquantifiable factors. These have not been quantitatively assessed or discussed in this book as they form separate topics by themselves. Nevertheless, their contributions form as important an aspect of warfare as the strategy adopted, the planning at the operational level, the tactics employed, and the relative or comparative sophistication of weapons and platforms in use. In fact, they may often make **the** difference between winning and losing.

ACRONYMS & SPECIAL TERMS SEQUENTIALLY

SONAR: SOund Navigation And Ranging.
SRP: Sound Ray Predictions. Predicting which path the transmitted or received sound ray is likely to take owing to varying sea conditions.
ESR: Expected Sonar Range. Maximum effective range under changing sea conditions for a given sonar transmission.
AIP: Air Independent Propulsion. Used by modern conventional submarines to increase underwater endurance and range without having to rely on air from above the surface of the sea.
SINS: Ship's Inertial Navigation System. An accurate underwater system that monitors the submarine's position from an established reference point in space.
SSN: Submersible Ship Nuclear/Nuclear Attack Submarine.
SR: Staff Requirements
SSK: Conventional Anti-submarine submarine.
DSRV: Deep Submergence Rescue Vehicle/Vessel. A small rescue submersible designed to dock with a sunken submarine and rescue the crew.
SSBN: Ship Submersible Ballistic Nuclear/ Strategic Missile Submarine, nuclear powered.
MRV: Multiple Re-entry Vehicle. Missile with more than one warhead
MIRV: Multiple Independently target-able Re-entry Vehicle. A Missile that has multiple heads, each targeting a different object.
SAG: Surface Action Group of ships.
SLBM: Submarine Launched Ballistic Missile.
SLCM: Submarine Launched Cruise Missile
SSGN: Nuclear-guided (cruise) missile submarine.
'Q' Tank: Sea Water Tank inside the submarine that rapidly takes in water to create negative buoyancy to go down quickly.
GRT: Gross Registered Tonnage of a vessel
SS-N-..: NATO assigned terminologies for Russian anti Surface /anti Submarine missiles with nuclear warheads launched by ships or submarines. The last number was just a differentiating serial number.
SOSUS: SOund SUrveillance System. Static Underwater Acoustic Sensors anchored to the seabed to detect incoming ships or submarines.
WMD: Weapons of Mass Destruction – nuclear, chemical, gaseous etc
ROE: Rules of Engagement for units at sea, and the airspace above.
HF/VHF/UHF: High Frequency/Very High Frequency/Ultra High Frequency radio waves.
SHF: Super High Frequency.
VLF: Very Low Frequency.

PQ17/PQ18/QP14: Numbers allocated to convoys during WWII..
IOR: Indian Ocean Region
SOA: Speed of Advance
LRMP: Long Range Maritime Patrol Aircraft
MRASW: Maritime Reconnaissance and Anti-Submarine Warfare Aircraft.
MOE: Measure of Effectiveness.
ASW: Anti-submarine Warfare
R/V: Rendezvous
CO: Commanding Officer. Also referred to as the 'Skipper', 'Captain', or the 'Old Man'.
PLA(N): The Naval element of the Chinese Armed Forces.
HMCS: Her Majesty's Canadian Ship.
C4SI: Command, Control, Communications, Computers, Space and Intelligence.
Indiscretion Rate: The ratio of the time needed to stay at Periscope Depth and the total operating time.
HMS: Her Majesty's Ship.
LLA: Limiting Lines of Approach. The angle these lines subtend between them increases or decreases based on the speed of advance of the submarine's likely target.
SURTASS: SURveillance Towed Array Sensor System. Something like a mobile SOSUS, towed by smaller Ocean Surveillance ships.
FCC: Fire Control Computer. Names differ in different navies.
HE: Hydrophone Effect. It is the noise made by propellers due to Cavitation.
ESM: Electronic Support Measures. It is a passive receiver system designed to track radar transmissions of aircraft and ships.
MFC: Missile Firing Computer. Like the FCC, different navies have different names for this.
VLS: Vertical Launch System. These are vertically placed in the submarine and fire missiles upwards as opposed to those that can be fired through torpedo tubes in lieu of torpedoes.
LOC: Lock Out Chambers. Used by Special Forces disembarked and recovered clandestinely by submarines while still dived.
UAV: Unmanned Airborne Vehicles.
UUV: Unmanned Underwater Vehicles.
SLOC: Sea Lanes of Communications. The shipping routes followed by merchantmen to ply cargo.
EHF MDR: Extreme High Frequency Medium Data Rate. Basically satellite communications.
ELF: Extreme Low Frequency.
SEAL: SEa-Air-Land. US Navy special force units/commandos.

ABOUT THE AUTHOR

Commodore P.R. Franklin, AVSM, VSM, (Retd) did his submarine training in Vladivostok, in erstwhile USSR. He commissioned the third and the sixth of the eight 'Foxtrot' class submarines that the Indian Navy purchased, and sailed them from Riga, Latvia, to India round Africa in the late sixties/early seventies. He subsequently commanded two of them. After a few squadron appointments both afloat and ashore, he headed the submarine arm as Director Submarine Operations in Naval Headquarters. Among his other appointments in the navy, he commanded the Training Squadron training officer cadets of the Indian Navy, INS Venduruthy in Kochi, was Naval Assistant to the Chief of the Naval Staff, the Naval Advisor to the Indian High Commissioner in the United Kingdom, and the Naval Officer-in-Charge Tamil Nadu & Pondicherry.

A graduate of the Defence Services Staff College, he also served as Directing Staff in that institution. He did the higher command and staff course in the former Marshal A.A.Grechkov Academy in Leningrad (renamed now as the N.G. Kuznetsov Academy in St. Petersburg). He was awarded the Vishisht Seva Medal in 1995 and the Ati Vishisht Seva Medal in 2001 by the President of India.

After retiring from the navy after 36 years of commissioned service, he was a consultant for a brief spell to a private Indian company while it produced a Submarine Control Simulator for the Indian nuclear propelled submarine, INS Arihant. He is also the author of the book titled 'Submarine Operations'.

His e-mail address is jalvayufranklin1946@gmail.com

Cmde P.R. Franklin, AVSM, VSM, (Retd)

Get Published with Frontier India

Do you want to get your book or thesis published? You might even want to republish your book which is currently out of print.. Frontier India Technology as a publisher, distributor and retailer of books, offers a complete range of publishing, editorial, and marketing services that helps you as an author to take his or her book to the reader.

Getting your work published is a wish for many for reasons including profit earning, self-satisfaction, popularity and other good reasons. We will offer you choices based on your needs. Get in touch with us at frontierindia@gmail.com.

Our Recently Published Books include :

An Indian Air force Recollects by Wing Co P.K. Karayi (Retd.) ISBN: 978-8193005507

Warring Navies – India and Pakistan (International Edition) – by Cmde Ranjit B. Rai (Retd.). Joseph P. Chacko. ISBN: 978-8193005545

Basics of marriage Management by Walter E Vieira. ISBN: 978-8193005514

Beat That Exam Fever – Succeed in Examinations by Walter E Vieira. ISBN: 978-8193005538

Ordinary Stocks, Extra Ordinary Profits by Anand S. ISBN: 978-8193005521

Foxtrot to Arihant – The Story of Indian Navy's Submarine Arm by Joseph P. Chacko. ISBN: 978-8193005552

Foxtrots of the Indian Navy by Cmde P.R Franklin. ISBN: 978-8193005576

A Nation and its Navy at War by Cmde Ranjit Rai. ISBN: 978-8193005583

The Role of the President of India by Prof Balakrishna. ISBN: 978-8193005569